Electrical Machine Principles: A Handbook

Electrical Machine Principles: A Handbook

Contributors

M. Benhaddadi, G. Olivier et al.

www.aurisreference.com

Electrical Machine Principles: A Handbook

Contributors: M. Benhaddadi, G. Olivier et al.

Published by Auris Reference Limited
www.aurisreference.com

United Kingdom

Copyright 2016
Printed in 2017 for Sale in the Indian Subcontinent

The information in this book has been obtained from highly regarded resources. The copyrights for individual articles remain with the authors, as indicated. All chapters are distributed under the terms of the Creative Commons Attribution License, which permit unrestricted use, distribution, and reproduction in any medium, provided the original author and source are credited.

Notice

Contributors, whose names have been given on the book cover, are not associated with the Publisher. The editors and the Publisher have attempted to trace the copyright holders of all material reproduced in this publication and apologise to copyright holders if permission has not been obtained. If any copyright holder has not been acknowledged, please write to us so we may rectify.

Reasonable efforts have been made to publish reliable data. The views articulated in the chapters are those of the individual contributors, and not necessarily those of the editors or the Publisher. Editors and/or the Publisher are not responsible for the accuracy of the information in the published chapters or consequences from their use. The Publisher accepts no responsibility for any damage or grievance to individual(s) or property arising out of the use of any material(s), instruction(s), methods or thoughts in the book.

Electrical Machine Principles: A Handbook

ISBN: 978-1-78154-921-6

British Library Cataloguing in Publication Data
A CIP record for this book is available from the British Library

Printed in the United Kingdom

Exclusively distributed by CBS Publishers & Distributors Pvt. Ltd.

Sales & Distribution Rights only for India, Pakistan, Bangladesh, Sri Lanka, Nepal and Bhutan. This book is not to be sold outside these territories.

Contents

List of Abbreviations vii
List of Contributors ix
Preface xiii

Chapter 1   **Premium Efficiency Motors** 1
M. Benhaddadi, G. Olivier, R. Ibtiouen, J. Yelle, and J-F Tremblay

Chapter 2   **A New Type of Capacitive Machine** 31
Arie Shenkman, Saad Tapuchi, and Dmitry Baimel

Chapter 3   **Optimal Design and Control of a Torque Motor for Machine Tools** 45
Yee-Pien YANG, Shih-Chin YANG, and Jieng-Jang LIU

Chapter 4   **Electric Motor Performance Improvement Using Auxiliary Windings and Capacitance Injection** 67
Nicolae D.V

Chapter 5   **Magnetic Reluctance Method for Dynamical Modeling of Squirrel Cage Induction Machines** 89
Jalal Nazarzadeh and Vahid Naeini

Chapter 6   **Minimization of Losses in Converterfed Induction Motors – Optimal Flux Solution** 111
Waldiberto de Lima Pires, Hugo Gustavo Gomez Mello, Sebastião Lauro Nau, and Alexandre Postól Sobrinho

Chapter 7   **Sensorless Vector Control of Induction Motor Drive - A Model Based Approach** 131
Jogendra Singh Thongam and Rachid Beguenane

Chapter 8   **Feedback Linearization of Speed-Sensorless Induction Motor Control with Torque Compensation** 161
Cristiane Cauduro Gastaldini, Rodrigo Zelir Azzolin, Rodrigo Padilha Vieira, and Hilton Abílio Gründling

Chapter 9   **A Rmrac Parameter Identification Algorithm Applied to Induction Machines** 181
Rodrigo Z. Azzolin, Cristiane C. Gastaldini, Rodrigo P. Vieira, and Hilton A. Gründling

Chapter 10	**Swarm Intelligence Based Controller for Electric Machines and Hybrid Electric Vehicles Applications** 201
	Omar Hegazy, Amr Amin, and Joeri Van Mierlo
Chapter 11	**Operation of Active Front-End Rectifier in Electric Drive Under Unbalanced Voltage Supply** 239
	Miroslav Chomat

Citations 271

Index 273

List of Abbreviations

ASD	Adjustable speed drive
IEEE	Institute of electrical and electronics engineers
IEC	International electrotechnical commission
MEPS	Minimum energy performance standards
AC	Alternative current
DC	Direct current
EMF	Electromotive force
PM	Permanent magnet
FEM	Finite element method
MECM	Magnetic equivalent circuit method
FLC	Feedback linearization control
IM	Induction motor
MRAS	Model reference adaptive system
FOC	Field-oriented control
MRC	Model reference control
RMRAC	Robust model reference adaptive controller
SPIM	Single-phase induction motors
FOC	Field-oriented controller
FCHEV	Fuel cell/supercapacitor hybrid electric vehicles
PSO	Particle swarm optimization
SI	Swarm intelligence
TPIM	Two-phase induction motor

List of Contributors

M. Benhaddadi
École Polytechnique de Montréal, dépt. de génie électrique, C.P. 6079 Succursale Centre-ville, Montréal, Québec, H3C 3A7 Canada

G. Olivier
École Polytechnique de Montréal, dépt. de génie électrique, C.P. 6079 Succursale Centre-ville, Montréal, Québec, H3C 3A7 Canada

R. Ibtiouen
École Nationale Polytechnique d'Alger, dépt. de génie électrique, Avenue Pasteur BP 182, El Harrach, 16200 Alger

J. Yelle
Cégep du Vieux Montréal, dépt. Technologie de génie électrique, 255 Ontario-Est Montréal, Québec, Canada H2X 1X6 Canada

J-F Tremblay
Cégep du Vieux Montréal, dépt. Technologie de génie électrique, 255 Ontario-Est Montréal, Québec, Canada H2X 1X6 Canada

Arie Shenkman
Electrical and Electronics Engineering Department, Shamoon College of Engineering, Beer Sheva, Israel

Saad Tapuchi
Electrical and Electronics Engineering Department, Shamoon College of Engineering, Beer Sheva, Israel

Dmitry Baimel
Electrical and Electronics Engineering Department, Shamoon College of Engineering, Beer Sheva, Israel

Yee-Pien YANG
Department of Mechanical Engineering, National Taiwan University, Taipei, Taiwan, China.

Shih-Chin YANG
Department of Mechanical Engineering, National Taiwan University, Taipei, Taiwan, China.

Jieng-Jang LIU
Department of Mechanical Engineering, National Taiwan University, Taipei, Taiwan, China.

Nicolae D.V
Tshwane University of Technology South Africa

Jalal Nazarzadeh
Faculty of Engineering Shahed University, Tehran Iran

Vahid Naeini
Faculty of Engineering Shahed University, Tehran Iran

Waldiberto de Lima Pires
WEG Equipamentos Eletricos S.A. – Motores Research and Development of Product Department Av. Pref. Waldemar Grubba, 3000 – malote 41 Jaraguá do Sul, SC - 89256-900 Brazil

Hugo Gustavo Gomez Mello
WEG Equipamentos Eletricos S.A. – Motores Research and Development of Product Department Av. Pref. Waldemar Grubba, 3000 – malote 41 Jaraguá do Sul, SC - 89256-900 Brazil

Sebastião Lauro Nau
WEG Equipamentos Eletricos S.A. – Motores Research and Development of Product Department Av. Pref. Waldemar Grubba, 3000 – malote 41 Jaraguá do Sul, SC - 89256-900 Brazil

Alexandre Postól Sobrinho
WEG Equipamentos Eletricos S.A. – Motores Research and Development of Product Department Av. Pref. Waldemar Grubba, 3000 – malote 41 Jaraguá do Sul, SC - 89256-900 Brazil

Jogendra Singh Thongam
Department of Renewable Energy Systems, STAS Inc., Chicoutimi, QC

Rachid Beguenane
Department of ECE, Royal Military College, Kingston, ON Canada

Cristiane Cauduro Gastaldini
Federal University of Santa Maria

Rodrigo Zelir Azzolin
Federal University of Santa Maria
Federal University of Rio Grande

Rodrigo Padilha Vieira
Federal University of Santa Maria
Federal University of Pampa Brazil

Hilton Abílio Gründling
Federal University of Santa Maria

Rodrigo Z. Azzolin
Federal University of Santa Maria
Federal University of Rio Grande

Cristiane C. Gastaldini
Federal University of Santa Maria

Rodrigo P. Vieira
Federal University of Santa Maria
Federal University of Pampa Brazil

Hilton A. Gründling
Federal University of Santa Maria

Omar Hegazy
Faculty of Engineering Sciences Department of ETEC- Vrije Universiteit Brussel, Belgium

Amr Amin
Power and Electrical Machines Department, Faculty of Engineering – Helwan University, Egypt

Joeri Van Mierlo
Faculty of Engineering Sciences Department of ETEC- Vrije Universiteit Brussel, Belgium

Miroslav Chomat
Institute of Thermomechanics AS CR, v.v.i. Czech Republic

Preface

Electric machine is synonymous with electric motor or electric generator, all of which are electromechanical energy converters: converting electricity to mechanical power (i.e., electric motor) or mechanical power to electricity (i.e., electric generator). The movement involved in the mechanical power can be rotating or linear. The apparatus that converts energy in three categories: generators which convert mechanical energy to electrical energy, motors which convert electrical energy to mechanical energy, and transformers which changes the voltage level of an alternating current. Developing ever more efficient electric machine technology and influencing their use are crucial to any global conservation, green energy, or alternative energy strategy. Electrical Machine Principles: A Handbook will give students and professionals alike all of the information essential to advance a full understanding of electrical machines in contemporary industry. First chapters illustrates the induced enormous energy saving potential, permitted by using high-efficiency motors. Furthermore, the most important barriers to larger high-efficiency motors utilization are identified, and some incentives recommendations are given to overcome identified impediments. Second chapter proposes a new type of the synchronous capacitive machine operated on a principle of the electric field effect. The proposed machine has smaller size and lighter weight than the standard electromagnetic synchronous machines with the same rated parameters. Another important advantage is a simple structure of the machine, which simplifies the production process and reduces the costs of the motor. The paper also presents extensive simulation results of the proposed capacitive machine. Third chapter presents a systematic approach of optimal design and control of a surface-mount, permanent-magnet synchronous torque motor for the next-generation machine tools. Fourth chapter explores on the electric motor performance improvement using auxiliary windings and capacitance injection. In fifth chapter, some classical techniques are introduced for identification and diagnosis of induction machines faults. Additionally, other heuristic methods are proposed to monitor of the induction machines for fault detection. minimization of losses in converter-FED induction motors – optimal flux solution are presented in sixth chapter. The aim of seventh chapter is to provide with a brief overview of high performance sensorless induction motor drive. Two sensorless vector control strategies using machine model-based estimation are presented in this chapter. Eighth chapter addresses the problem of controlling a three-phase Induction Motor (IM) without mechanical sensor (i.e. speed, position or torque measurements). The purpose of this chapter is to present the development of two FLC control strategies in the presence of torque disturbance or load variation, especially under low rotor speed conditions. Ninth chapter deals with the problem of parameter identification of electrical machines to achieve good performance of a control system. In tenth chapter, a field-oriented controller that is based on Particle Swarm Optimization is presented. In this system, the speed control of two asymmetrical windings induction motor is achieved while maintaining maximum efficiency of the motor. Last chapter deals with the effects of unbalanced voltage supply on the DC-link voltage pulsations, methods to address this problem and the additionally imposed constraints in operating regions of the rectifier.

Chapter 1

PREMIUM EFFICIENCY MOTORS

M. Benhaddadi[1], G. Olivier[1], R. Ibtiouen[2], J. Yelle[3], and J-F Tremblay[3]

[1]École Polytechnique de Montréal, dépt. de génie électrique, C.P. 6079 Succursale Centre-ville, Montréal, Québec, H3C 3A7 Canada

[2]École Nationale Polytechnique d'Alger, dépt. de génie électrique, Avenue Pasteur BP 182, El Harrach, 16200 Alger

[3]Cégep du Vieux Montréal, dépt. Technologie de génie électrique, 255 Ontario-Est Montréal, Québec, Canada H2X 1X6 Canada

INTRODUCTION

Despite its considerable potential for energy savings, energy efficiency is still far from realizing this potential. This is particularly true in the electrical sector (IEA, 2010). Why? There is no probably just one single answer to this question. A consequential response requires major multiform research and an analytical effort. No doubt that analysis of the interaction between energy efficiency policies and energy efficiency performance of economies accounts for a significant part of the effort.

In the future sustainable energy mix, a key role will be reserved for electricity, as GHG emissions reduction in this sector has to be drastically reduced. In this option, obvious conclusion is that large market penetration Premium motors needs a complex approach with a combination of financial incentives and mandatory legal actions, as industry doesn't invest according to least life cycle costs (DOE, 2010).

This present work illustrates the induced enormous energy saving potential, permitted by using high-efficiency motors. Furthermore, the most important barriers to larger high-efficiency motors utilization are identified, and some incentives recommendations are given to overcome identified impediments.

In the present work, experimental comparison of the performance characteristics of 3 hp Premium efficiency motors from three different manufacturers has been presented. The motors were tested according to Standard IEEE 112-B.

ENERGY, CLIMATE CHANGE AND ELECTRICITY

According to last report Intergovernmental Panel on Climate Change IPCC report (IPCC, 2007), the observed increase in global average temperatures since the mid-20th century is very likely due to the observed increase in anthropogenic greenhouse gas concentrations. Moreover, there is no doubt that discernible human influences now extend to other aspects of climate, including ocean warming, continental-average temperatures, temperature extremes and wind patterns. Stabilizing atmospheric carbon dioxide concentrations at twice the level of pre-industrial times is likely to require emissions reductions up to 90 % below current levels by 2100. Clearly, reductions of this magnitude can be achieved only by taking action globally and across all sectors of the economy. The electricity sector will undoubtedly need to assume a major share of the weight, according to its contribution to overall emissions estimated to be more 10 000 Mt (million tone) CO_{2eq} per year.

As can be seen in fig.1, the electricity generation is dominantly produced from fossil fuels (coal, oil, and gas), and today's situation is the same as forty years ago (DOE, 2010). In the last XXI world energy congress, it is highlighted that electricity generation will still depend on fossil sources. In the meantime, according to (IEA, 2010), industry accounts for more 40 % of the world 20 000 TWh (terawatt hours, or so called billion kilowatt hours) electricity consumption, weighting more 4 000 Mt CO_{2eq} per year. Within the industrial sector, motor driven systems account for approximately 60% to 65% of the electricity consumed by North American (RNC 2004, DOE 2010) and European Union industries. Implementing high efficiency motor driven systems, or improving existing ones just by 1 to 2 %, could save up to 100-200 TWh of electricity per year. This would significantly reduce the need for new power plants. It would also reduce the production of greenhouse gases by more 100 million CO_{2eq} per year and push down the total environmental cost of electricity generation.

The worldwide electric motors above 1 hp can be estimated to be nowadays more 300 million units, with the annual sales of 34 million pieces. Typically, one-third of the electrical energy use in the commercial sector and two-thirds of the industrial sector feed the electrical motors (DOE, 2010). Moreover, the low voltage squirrel cage induction motor constitutes the industry workhorse. In particular industrial sector such as the Canadian petroleum and paper industry, the share of the energy used by electrical motors can reach 90% (RNC 2004). Since induction motors are the largest electrical energy user, even small efficiency improvements will result in very large energy savings and contribute to reduce greenhouse gas emissions GHG. Furthermore, the declining resources combined environmental global warming concerns and with increasing energy prices make energy efficiency an imperative objective.

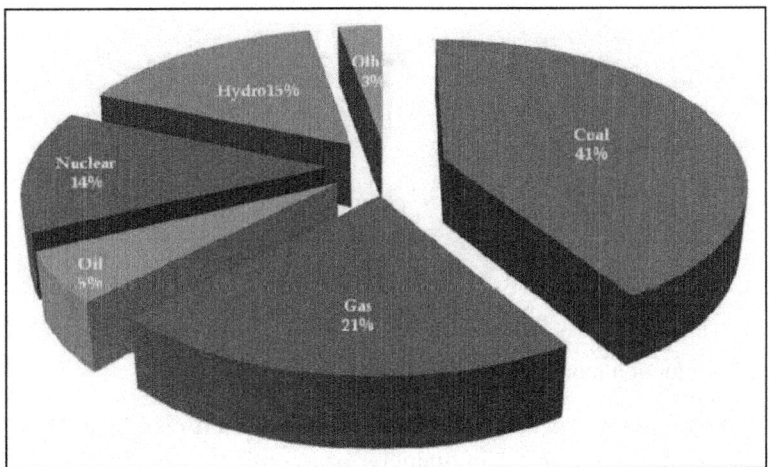

Figure 1: Electricity generation by fuel

MOTOR LOSSES SEGREGATION AND EFFICIENCY

The impact of a motor in terms of energy and economical costs depends on its performance during its lifetime. The motor performances are characterized by the efficiency with which it converts electrical energy into mechanical energy.

In Standard IEEE 112-B the losses are segregated and the efficiency is estimated by the following formula:

$$\Delta P_{str} = P_{in} - P_{out} - (\Delta P_{el1} + \Delta P_{el2} + \Delta P_{core} + \Delta P_{mech}) \quad (1)$$

Where the electric input power, P_{in}, is measured with a power analyser and the output power, P_{out}, with a torque meter. The overall precision of efficiency assessment mainly depends on the torque estimation, and with the improved accuracy of recent power analysers and torque meters, this method can be considered accurate and reliable.

Motor efficiency is defined as a ratio motor mechanical output power and electrical input power. Hence in order to have a motor perform better, it is important to reduce its losses. The major motor losses are resistive losses in the stator and the rotor windings, and magnetic losses (hysteresis and eddy current losses) in the cores. Other losses include mechanical (bearing friction and ventilation), and stray load losses. High efficiency motor losses relative distribution is not so different at low efficiency one's; it's more dependent on the power. Their general distribution is illustrated in fig.2.

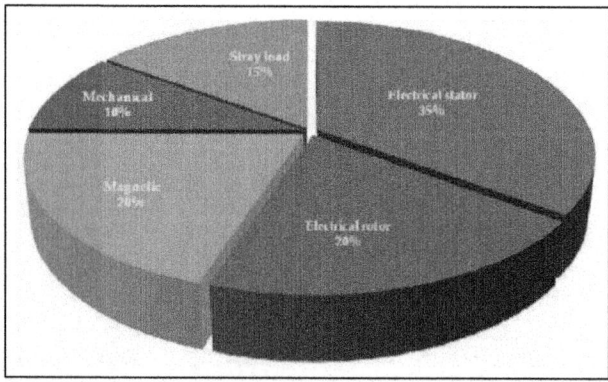

Figure 2: Induction motor losses distribution

There are many ways to improve electric motor efficiency; the majority of them make the motor larger in diameter or overall sizes and, of course, more expensive.

- Winding stator (ΔP_{el1}) and rotor (ΔP_{el2}) losses are due to currents flowing through the stator windings and rotor bars. These losses can be reduced by decreasing the conductor current density in the stator windings, in the rotor bars and in the end rings. Using larger conductors lowers stator resistance, while the use of copper instead of aluminum reduces rotor losses (Parasiliti et al. 2002). Another way of decreasing stator losses is by reducing the number of turns. Unfortunately, this increases the starting current and maximum torque, as worsen the power factor.
- Magnetic losses ΔP_m occurring in the stator and rotor laminations are caused by the hysteresis and eddy current phenomena. These losses can be decreased by using better grade magnetic steel, thinner laminations and by lowering the flux density (i.e larger magnetic cores). The better grade of laminations steel are still relatively very expensive. Cheaper manufacturing methods other than stamping are expected to become available in the near future.
- Mechanical losses ΔP_{mec} are due to bearing friction and cooling fan air resistance. Improving the fan efficiency, the air flow and using low friction bearings result in a more efficient design. As these losses are relatively small, the efficiency gain is small too, but every improvement is welcome.
- Stray load losses ΔP_{st} are due to leakage fluxes induced by load current,

non-uniform current distribution, mechanical air-gap imperfection... These losses can be reduced by design optimisation and manufacturing method improvements.

As can be deducted, one of the most established methods of increasing motor efficiency is to use higher quality materials, inexorably increasing the motor cost, as most high performance materials are expensive materials. In a recurrent manner, the same problem of increased cost holds true for better construction techniques, such as smaller air gaps, copper rather than aluminum in the rotor construction, higher conductor slot fill, and segmented core stator construction. The resulting increase in motor cost is evaluated to be between 15 % and 30 %.

TESTING STANDARDS

In North America, the prevailing testing method is based on direct efficiency measurement method, as described in the Institute of Electrical and Electronics Engineers (IEEE) *"Standard Test Procedure for Polyphase Induction Motors and Generators"* IEEE 112-B and in its Canadian CSA 390 adaptation. The standard first introduced in 1984 and updated in 2004, requires the measurement of the mechanical power output and the electric input, and provide a value for the motor losses, where the additional stray load losses are extrapolated from their total by the following formula (1). So, the efficiency is extrapolated by:

$$\eta = \frac{Pout}{Pin} = \frac{Pout}{Pout + \Delta Pel1 + \Delta Pel2 + \Delta Pm + \Delta Pmec + \Delta Pst} \qquad (2)$$

In Europe, the prevailing testing method is based on an indirect efficiency measurement as defined in IEC 34-2 standard "Rotating electrical machines – Part 2: Methods for determining losses and efficiency of rotating electrical machinery from tests". The standard first introduced in 1972 and updated in 1997, attribute a fixed value, equal to 0.5 % of input power to the additional stray load losses.

These standards differ mainly by the method used to take into account the additional load losses (Aoulkadi & Binder, 2008, Boglietti et al 2004, Nagorny et al. 2004, Elmeida et al. 2002...). Many papers have been published and some authors have illustrated, that IEC 34 – 2 has drawback with a noticeable influence on the testing of high efficiency motors, as the efficiency of this motor type is overestimated, particularly in the small motor size cases. Ultimately, standard IEC 34 – 2 was found to be unrealistic with its 0.5 % P_{in} value for stray losses (Aoulkadi & Binder, 2008, Renier et al. 1999, Boglietti et al 2004...). That is why, in 2007, IEC published a revised standard for efficiency classification no. 60034-2-1 which includes a test procedure largely comparable to IEEE 112-B or CSA C390. Newly harmonized standards for energy efficiency testing IEC 60034-2-1 can contribute to lowering barriers in global trade for energy efficient motor systems.

MINIMUM ENERGY PERFORMANCE STANDARD MEPS AND EFFICIENCY MOTOR CLASSIFICATION

There are many different worldwide definitions for energy efficient motors, as until these last years, there was no consensus on what really represents an energy efficient motor. Technical barriers include nonharmonized testing standards and efficiency classification. In reality, the key mandatory instrument is minimum energy performance standards (MEPS).

MEPS in North America

On October 1992, US Congress voted law, Energy Policy Act EPAct, which mandates strict energy efficiency standards for electrical appliances and equipment, including electric motors. Motor MEPS were for the first time introduced in 1992 when all partners were finally persuaded that voluntary measures are too slow, and no significant market transformation towards more efficient motors was possible otherwise.

EPAct requires that the general purpose electric motors meet the higher nominal efficiency requirements defined in the table of National manufacturer association NEMA Standard, and the implementation of the motor MEPS went into effect in 1997.

The Canadian Standard association developed a Canadian standard in 1993, and updated it in 1998. CAN/CSA C-390 set the requirement for minimum efficiency for new motors made or sold in Canada at the same value as the NEMA energy-efficient level.

The Energy Policy Act EPAct-92 motors covered are:
- General purpose
- Definite or special purpose in a general purpose application
- Continuous duty
- 2, 4 & 6 Pole
- 1-200 HP
- 230/460/, 60 Hz

The Canadian standard was furthermore extended to 575 V and IEC motors, and included

75 % full load to reach maximum efficiency.

Some motors were not covered:
- Definite or special purpose in a non-general purpose application
- Slower speeds
- Inverter duty
- Multi-speed
- Totally enclosed air over TEAO, and totally enclosed not ventilated TENV

As a result of the mandatory standard that was endorsed as part of the EPAct-92, North America had a motor standard foundation that leads the new century world.

In 2002, NEMA and Consortium for energy efficiency CEE established a voluntary NEMA Premium level of efficiency, and the manufacturers began the next step in evolution with the implementation on voluntary basis MEPS NEMA Premium efficiency motors. NEMA premium efficiency standards (CEE 2007) have remained voluntary for a long period of 10 years. In spite of this, NEMA premium motors have been progressively gaining market share, as the overall benefits of Premium motors is incommensurable (more reliable, last longer, have longer warranties, run more quietly and cooler and produce less waste heat than their less-efficient counterparts). The trend is particularly well depicted in the work, illustrated in fig. 3.

The evolution of MEPS based on NEMA Premium is now moving from voluntary basis to legislated regulation, as the law implementation is awaited for December 2010. So, 1–200 HP general purpose motors already covered by EPAct will change from NEMA MG-1, Table 12-11 Energy Efficient (Annex 1) to Table 12-12 NEMA Premium efficiencies levels (Annex 2), except for fire pump motors which remain at EPAct-92 level.

Moreover, the proposal expands the scope of enclosed 1-200 Hp motors, as several motor types not previously covered by EPAct-92 must meet EPAct efficiency levels. The added motors are:
- U-frame
- Design C
- Footless
- Close-coupled pump
- Vertical solid shaft normal thrust
- 8 poles
- All low voltages ≤ 600 V not previously covered, including IEC Metric frame motors from 90 frames and up.

- 201 to 500 HP low voltage 2-8 pole general purpose motors, where Design B represent something like ¾ of total range.

It's appropriate to highlight that this additional coverage means that MEPS is extended to more 90 %of the motors in the 1 to 200 Hp range, despite the fact that some manufacturers expressed technical skepticism about meeting premium efficiency levels design C and 900 rpm motors.

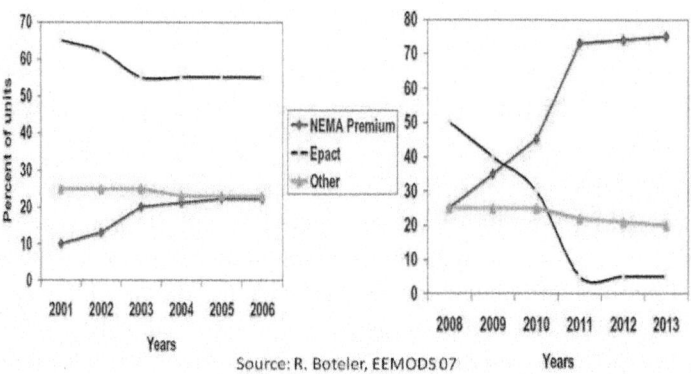

Source: R. Boteler, EEMODS 07

Figure 3: Efficiency trend 2001-2006 and efficiency expected 2008-2013

On the other hand, EISA does not apply to motors exported outside North America, as well as EISA doesn't contain any requirement to replace electric motors in use. In addition, the law applies only to motors manufactured after December 19, 2010 and motors in inventory on that date can be sold or used without any additional constraints, but some customers may not probably want them.

Some Motor configurations are not covered by EISA law. These are:
- Design D with high slip
- Inverter duty with optimized windings for adjustable speed drives ASD use
- Customized OEM mounting
- Intermittent duty
- Integral with gearing or brake where motor cannot be used separately
- Submersible motors
- Single Phase motors
- DC motors
- Two-digit frames (48-56)
- Multi-speed

- Medium voltage motors
- Repaired motors
- TENV and TEAO enclosures

It's relevant to notice that regulation concerning single phase and three-phase small (42/48 and 56 frame) motors is ready, as Department of energy (DOE 2010) published in 2010 its final rule giving efficiency limits that must be meet by small general purpose electric motors, starting from 2015. Nevertheless, NEMA doesn't support regulation of this product, as many questions are still not solved while small induction motor is continually losing market part to switched reluctance and permanent magnet types.

New IEC standard

In 2007, IEC published a revised standard for efficiency testing no. 60034-2-1 which includes a test procedure largely comparable to IEEE 112-B or CSA C390, and in September 2008 a new standard no. 60034-30 for efficiency classification of electric motors.

New standard for efficiency classification is applicable to single speed, three-phase induction motors with the following parameters:
- Power from 0.75 to 375 kW
- Voltage under 1000 V
- 50 and 60 Hz frequency
- 2, 4 and 6 poles
- Duty S1 or S3
- All IP1x to 6x and IC0x to 4x
- Networked
- All types of fixing, shaft extension, accessories

Some Motor configurations are not covered by IEC 60034-30. These are:
- Motors with reinforced isolation specifically designed for variable speed drives applications
- Motors which are fully integrated in a machine and cannot be tested separately

Newly 2008 harmonized standard for energy efficiency class 60034-30-2008 follow International Electrotechnical Commission (IEC) protocol and defines four induction motor efficiency classes:
- Super Premium efficiency level IE4
- Premium efficiency level IE3

- High efficiency level IE2
- Standard efficiency level IE1

It's noticeable that, nowadays, Premium efficiency IE3 is the most efficient motor. Super Premium efficiency IE4 is a future new generation motor. It is awaited that in average, the losses reduction of IE4 should be 15 % compared to IE3. So, IE4 is not a standard in fact, but just a level.

MEPS in EU and BRIC countries

In Europe, the European committee of manufacturers of electrical machines and power electronics CEMEP has classified 2 & 4 pole 1-90 kW motors into three levels:
- High (EFF1),
- Improved (EFF2), and
- Standard efficiency (EFF3).

The CEMEP classification has induced substantial EFF3 motors reduction and EFF2 market share promotion. Nevertheless, EFF1 market part is still modest. Meanwhile, European Union is considering prohibiting the sale of motors that don't meet EFF2 criteria in the near 2011 future.

The new EU MEPS system is based on the latest IEC standard, and it represents a significant step towards worldwide harmonization of efficiency regulations.

The challenges that EU countries facing are:
- From June 16, 2011, motors shall not be less efficient than IE1 (i.e. EFF2) and EFF3 motors will be banned
- From January 1, 2015, motors with rated output of 7.5 to 375 kW shall not be less efficient than IE2
- From January 1, 2017, motors with rated output of 7.5 to 375 kW shall not be less efficient than IE3
- From January 1, 2017, all motors with rated 0.75 to 375 kW shall not be less efficient than IE2 if equipped with adjustable speed drive (ASD).

BRIC countries (Brazil-Russia-India-China) motors and motor driven equipment are still relatively less efficient. For example, the efficiency of over 80 % Chinese motors is 2-5 % lower than international advanced ones. Chinese scientists consider that if efficiency of motor systems could be raised to the North American level, then 150 TWh of electricity would be saved each year. But China is making progress, by formulating a number of policies, laws and regulations on energy conservation. Beginning this 2010 year, the MEPS efficiency of newly added motors should reach international first-rate level Class 2, which is equivalent to IE1, or EFF2.

At the present time, 10 countries with a half of global electricity demand have motor MEPS endorsement. In the next two years, 15 new countries targeted for a next round until 2012 to reach around 80% of global electricity demand.

As earlier mentioned, the International Electro technical Commission (IEC), a worldwide organization for standardization comprising all national standards committees, defined in IEC 60034-30 (2009) four efficiency classes for single-speed cage-induction motors, and specified test procedures. It's important to notice that the so-called IE4 Super Premium Efficiency products are not commercially available yet, while lower efficiency motors in use now Eff3 disappears in the new classification. This new classification will be probably soon adopted worldwide in place of regional or local classification, as illustrated in table 1. This new standard defines efficiency classes and their containing minimum values (conditions).

Table 1: International motor efficiency classification

Efficiency Class	IEC	USA/Canada	CEMEP	China
Super Premium efficiency	IE4	-	-	-
Premium efficiency	IE3	NEMA Premium	-	-
High efficiency	IE2	EPAct	EFF1	Class 1
Standard efficiency	IE1	-	EFF2	Class 2
Below standard efficiency	-	-	EFF3	Class 2

LIFE CYCLE COST PREMIUM MOTORS

An electric motor is somewhat cheap to buy, but expensive to run. For example, a 3 hp Premium efficiency motor functioning 6 000 hours per year consumes about 1000 $ of electricity at $0.07/kWh. The purchase price for such a motor is about 500 $ and over the motor's 15-year life, the acquisition price represents only 3 % of the lifetime costs, while the cost of electricity accounts for 97 %. Finally, a 2 % increase in Premium motor efficiency over EFF1 translates in energy savings over that time nearly twice the cost difference. In addition, with a larger motor, the saving potential will be larger, and therefore payback periods would be shortened. For the 100 Hp motor, the acquisition price represents only 1 % of the lifetime costs, while the cost of electricity accounts for 99 %!!! Fig. 4 depicts typical lifetime cycle cost motor in the conservative case (Benhaddadi & Olivier, 2010a).

The average life cycle of the small power motors is of the order of 15 years, i.e. the equivalent of the average car range. The fundamental difference is in the fact that during this period, the cost of the electricity will represent 97 % of the cost of useful life cycle of the electric motor, while for the car motor, it represents only 10 %. Moreover, the car's internal combustion motors can rarely overcome 50 % efficiency, with an enormous negative impact to be paid in environmental pollution. We can deduct from this fact that the improvement of 1 % of the electrical motor efficiency will have the same impact as the reduction of the 10 % gas consumption car.

Moreover, the Canadian electricity costs are presently up to two times cheaper than elsewhere and high electricity prices reduce payback period. In addition, some Canadian utility companies and public agencies like Hydro Québec in Québec offer rebate programs to encourage customers to upgrade their standard motors to Premium efficiency (Benhaddadi & Olivier, 2010b). For motors from 1 to 75 hp, this program allows 600 $/hp to the customer and 150 $/hp to the distributor for each saved hp. Unfortunately, as a consequence of the lack of energy saving importance, the purchase of a new motor, as well as the rewinding of defective standard-efficiency motors, the choice of the motor is often driven by short term investment considerations, not on the cost of the electricity which can be saved.

The first law for energy efficient motors is the Energy Policy Act (EPAct) which mandates strict energy efficiency standards for electrical appliances and equipment. This law was first adopted in USA and became effective in Canada with the adoption of Standard CAN/CSA- C390-98. Today more than 75 % of the motors sold in North America are Premium efficiency and EPAct machines. This clearly indicates the positive effect of the energy law. In light of the above, and taking into consideration the very slow market transformation with just voluntary and incentive measures, there's no doubt that Premium motors will monopolize the dominant part of the market in the near 2013 future. This is well illustrated in fig. 3. There is no doubt that the appropriate legislation is the best way of achieving that goal (Benhaddadi & Olivier, 2009b). Only the latest energy efficient motor technologies should be manufactured and used. In general terms, North America is not on the leading edge for energy saving and conservation. Motor efficiency is an exception that should be at least maintained by EISA law implementation.

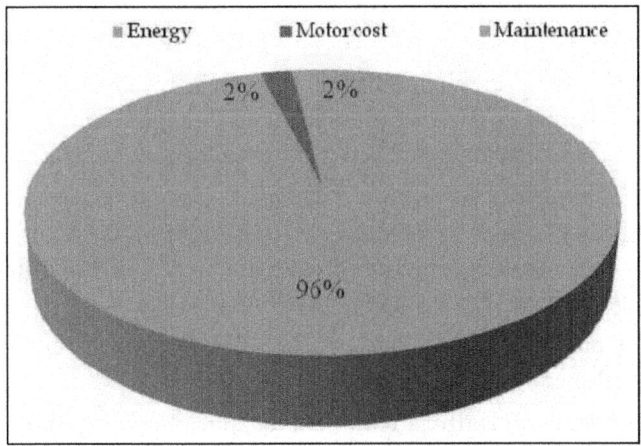

Figure 4: Lifetime motor cost

BARRIERS TO HIGH EFFICIENCY MOTORS MARKET PENETRATION

Despite the colossal energy saving potential and financial incentives programs, many companies are still reticent to invest in energy-efficiency motors. The reasons why the well-known potential for energy saving energy is not exploited have been investigated, and the authors have identified several reasons why this potential is not yet fully exploited. The Grand paradox is that cost effective measures are not taken because of several illogical barriers. So, the most important barriers to high efficiency motors promotion are (Benhaddadi & Olivier, 2008a):

- The energy costs are relatively so small that energy efficiency improvement isn't taken into consideration,
- Lower priority of energy savings importance, when other factors such as availability service, reliability, and first costs are of premium importance,
- Industry reluctance to change what is, a priori, a good functioning system,
- Doubt about success of energy efficiency programs, or the discount rates used to justify energy efficiency programs are too low,
- Downtime replacement cost look like peanuts, but shutdown time to install new equipment is expensive and many companies don't accept this inconvenient,

- Reduced budget often makes reducing energy consumption as « poor parent », inducing a lack of encouragement to make a decision,
- Implementing making-decision responsibility is often shared with many internal conflicting pressures and divergence and ultimate choice don't always belong to electrician engineers, who are energy savings conscious,
- Distributors regularly represent two or more motor manufacturers and they can advantage products from the manufacturer that offers the highest discount rather than high-efficiency ones,
- Usual predisposition to use stocked old motors rather than purchase high efficiency ones,
- It is not economically pragmatic to change a motor until it fails,
- Penchant to have the failed motor repaired rather than replaced by high efficiency ones,
- Degradation efficiency of repaired motor cannot be simply illustrated, Annual running hours are not sufficiently high to induce satisfactory payback.

INCENTIVE POLICIES TO OVERCOME BARRIERS

Experience derived from many energy saving initiatives around the world showed that the most successful programs are based on a combination of technical and promotional information, educational tools and financial incentives. If technicians and engineers would be trained in system design integration and least lifecycle cost as a goal, no doubt that the problem of inefficient industrial equipment should be solved (Benhaddadi & Olivier, 2008a). Consequently, to overcome the identified obstacles need a combination of the following measures:

- Premium priority: For the companies, the energy saving status has to arrive at legislative endorsement, like is the case for safety and quality insurance,
- Incentive programs: to reinforce energy savings promotion politics, much higher discount rates should be used to evaluate the cost-effectiveness of energy efficiency policies, programs or measures,
- Highlighted information and diffusion: this information must be of practical value, and sufficiently demonstrative with real pilot projects,
- Environmental concern: It's necessary to reinforce ecological policy criteria and support environmental friendly companies. This follows the principle that the saved energy is the most environmentally friendly one.

A particularly promising concept is the emissions trading scheme, which could be enable companies to claim emissions credits for investments that reduce energy consumption,
- Legislation: to legislate against recalcitrant and to impose to market the gradual approach of the « carrot and the stick », where the carrot represents the incentives and the stick stands for refractory.

The authors strongly believed in the need to enhance policy measures aimed at reducing the demand for energy and the resultant environmental impact. We therefore welcome the increased interest in energy conservation in Canada. But, the accumulated experience clearly show that putting in place incentives and voluntary measures for the energy efficiency of electric motors is not sufficient, as it is prerequisite to implement mandatory measures for larger market penetration (fig. 5). The carbon savings from this measure have the potential to make a significant contribution to emissions target reduction.

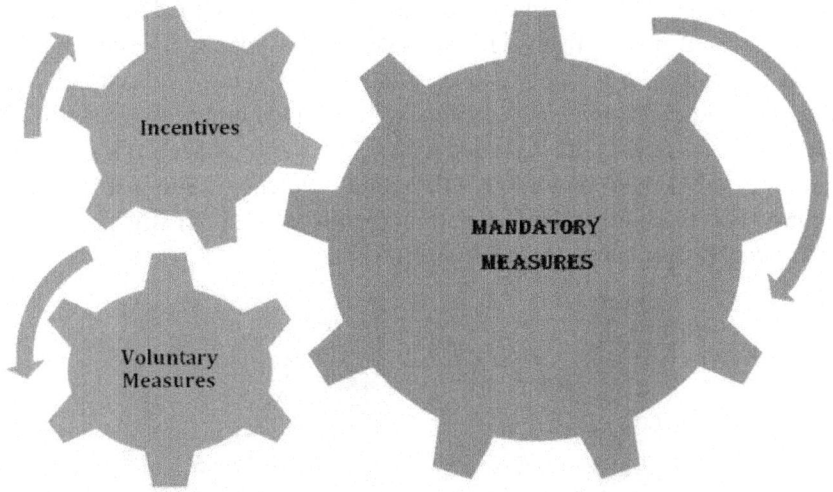

Figure 5: Incentives, voluntary and mandatory measures impact

EXPERIMENTAL SETUP AND RESULTS

Figure 6 depicts the experimental laboratory set-up. The motors are fed by three single phase autotransformers and direct torque control, DTC drives. The motors are mechanically loaded by dc machine connected through a precision torque-speed transducer. A motor/harmonic power meter is used for measuring real and reactive power, currents, voltages and power factor. The motors are mechanically loaded by dc machine connected through a high precision torque-speed transducer.

The measurements were taken in similar conditions and each motor was loaded until thermal equilibrium was reached, while each of the two benches can be used to determine the efficiency.

The two benches are sufficiently flexible and require minimum adjustments when different Premium motors are tested (Benhaddadi et al. 2010b). Several 3 hp Premium motors from different 3 manufacturers were tested. Figures 7 and 8 show the stators and rotors of the three different motors.

The measurements were taken in similar conditions and each motor was loaded until thermal equilibrium was reached.

For each of the motors B & C, Fig. 9 shows the variation of the efficiency versus the mechanical torque in the case of direct 60 Hz feeding, without drive. The results for motor A are not provided, as they are the same as for the motor B. One can deduct that when the motors are feed from the rated 230 V grid, the difference between the measured efficiency at rated load and the nameplate value is less a 0.2 %,. But, most electric motors are designed to function at 50 % to 100 % of rated value. Fig. 9 also shows that maximum Premium motor efficiency is near 75 % of rated load, and tends to decrease substantially below about 50 % load. Moreover, experienced overloaded motors don't significantly lose efficiency, as they are designed with a 1.15 factor service. It's also relevant to notice that the two motors give the same efficiency value just for the rated regime, as this efficiency difference is significant (1.5 %) when the motor is under loaded (Benhaddadi et al. 2010b).

Figure 6: Measurement setup with instrumentation

Next step is to analyze the feeding voltage impact. The voltages choice are made taking into account practical considerations: 230V, and 200.

- The first one is the motor nameplate indication, which is in accordance with an industrial available 240V feeding voltage,
- The second one is the real full-time laboratory available voltage. These two voltages were obtained with three one-phase transformers.

The experimental results obtained in the grid feeding case (without drive) and illustrated in fig. 10 show that feeding voltage has an important impact on efficiency value. With rated 230 V voltage and load values, efficiency is 89.3 %, i.e. 0.2 % less than nameplate value. When the feeding voltage is decreased to 208 V and 200 V values, the efficiency decrease up to 2 %. So, additional losses occur when a 230 V motor is operated at or below 208 volts. The motor show lower full-load efficiency, slips more, and produces less torque.

In another hand, the ASD (adjustable speed drive) deployment to control motor can substantially reduce energy consumption. The ASD advantages and energy consumption reduction are nowadays well documented (Benhaddadi & Olivier, 2007). But, at the present time, many companies use ASD to feed their motors, whenever they don't need speed regulation. In the energy saving point of view, we must be careful with this making-decision, as ASD introduces supplementary losses in the motor and the drive. As can be seen in fig. 11 and 12, the drive introduces a noticeable reduction of the motor system efficiency. This decrease reaches approximately 4 % in the rated regime, withdrawing totally energy savings induced by Premium efficiency motor use. Further investigations to correctly understand the extents of the losses introduced by the controller for all frequencies are under consideration.

Figure 7: Stator Premium motors

Figure 8: Rotor Premium motors

The other problem is that if Premium motors are misapplied, they may not achieve predicted energy savings and may result in diminished performance efficiency. For centrifugal pumps, an increase in operating speed will increase the required power by the third power of the speed ratio.

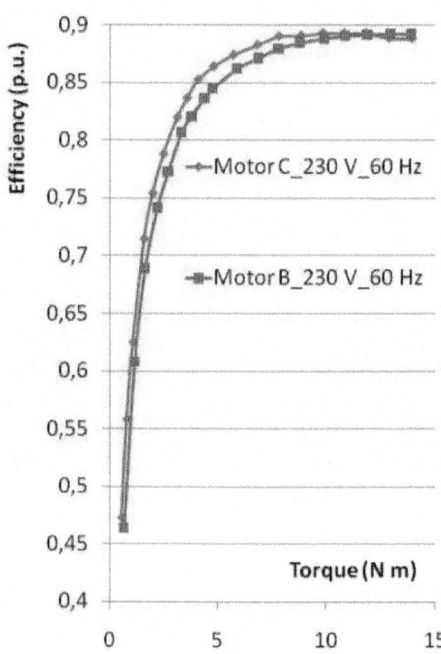

Figure 9: Manufacturing technology impact

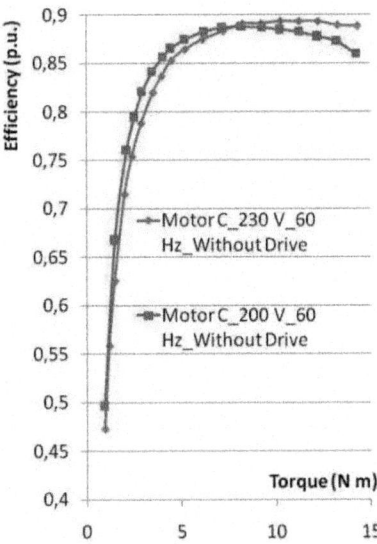

Figure 10: Viltage feeding impact

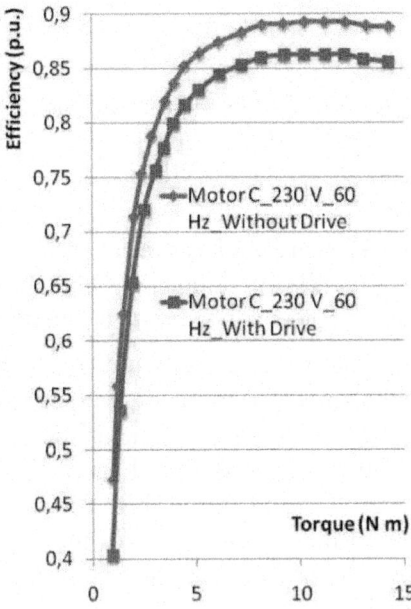

Figure 11: Drive impact

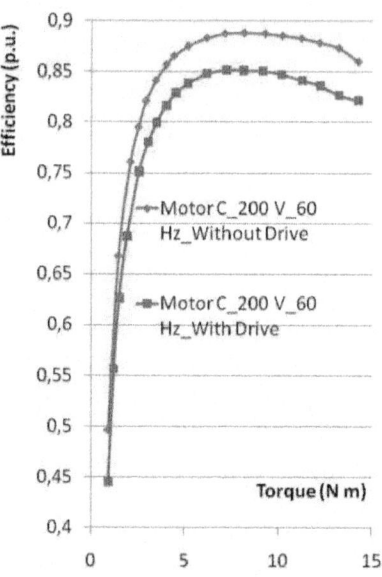

Figure 12: Drive impact

For example, by substituting Premium 1760 rpm motor to EPAct 1740 rpm one, a 20 rpm increase in the speed induce 3.5 % increase in the load, as $(1760 \div 1740)^3 = (1.014)^3 = 1.035$. So, when replacing a standard efficiency motor, one must be careful, as a Premium motor with lower or equal full-load speed must be selected to avoid the energy increase that may negate the predicted energy savings resulting from a higher efficiency.

It's important to notice that to date, there is no agreement that allows the determination of ASD system efficiency at any given frequency. The ideal situation is to obtain a family of efficiency curves for diverse torques and frequencies, including overrated values, as experimentally illustrated in fig. 13.

But, to generalize results, there are two difficulties: the first one can be illustrated by the results presented in fig.14, where we can see that the same 3 Hp motors issued from two different manufacturers can have the same efficiency for 60 Hz frequency feeding voltage, but different efficiency for another 30 Hz frequency (fig. 14). For each of these two motors B & C, Fig. 14 shows the variation of the efficiency versus the mechanical torque. In the 60 Hz case, the difference between the measured efficiency at rated load is negligible, while it reaches 2 % in the 30 Hz feeding frequency.

The second difficulty is about 50-60 Hz feeding frequency dilemma. As earlier illustrated, 50 Hz frequency gives better efficiency beginning from 2/3 load, while 60 Hz is better for low loads. It's noticeable that the same results were obtained for Motor B. Considering that for the same frequency, two Premium motors issued from two different manufacturers can show significantly different efficiency in low frequencies, one must be careful in generalizing the conclusions of this research. Moreover, the same motor tested for different frequencies can yield to different losses repartitions. So, before claiming that the obtained results are, or are not in agreement with findings of other authors, several other Premium efficiency motors from different constructors should be tested. The mentioned work is under consideration.

CONCLUSIONS

In the future sustainable energy mix, a key role will be reserved for electricity, as GHG emissions reduction in this sector has to be drastically reduced. In this option, obvious conclusion is that large market penetration Premium motors needs a complex approach with a combination of financial incentives and mandatory legal actions, as industry doesn't invest according to least life cycle costs.

The US Energy Policy Act and the Canadian Energy Efficient Act, along with the implementation of NEMA Premium efficiency levels, have lead to North American leadership on motor efficiency implementation. In general terms, North America is not on the leading edge for energy saving and conservation. Motor efficiency is an exception that should be at least maintained. Next step is to get Tax incentives to promote early retirement of older inefficient pre-EPAct motors by replacing instead of repairing

Experimental comparison of the performance characteristics of 3 hp Premium efficiency induction motors has been presented. The motors were tested according to Standard IEEE 112-B. In the rated frequency and voltage case, the experimental results are in good agreement with nameplate manufacturer's information. Particularly, a comparison of the rated operating point shows that, the discrepancy is approximately 0.2 %.

However, in low voltage/frequency applications, the use of a variable speed drive introduces extra losses and the overall efficiency can be noticeably reduced. The experimental results show that feeding voltage has an important impact on efficiency value, while efficiency at low frequencies depends on a certain level at manufacturer technology. From a global energy saving point of view, the ASD application to Premium efficiency motors should be promoted just when adjustable speed is needed.

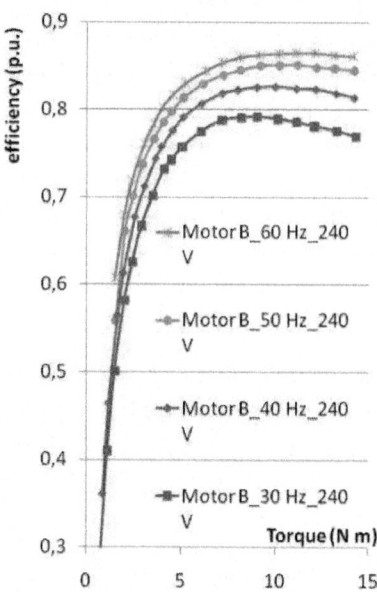

Figure 13: Frequency impact

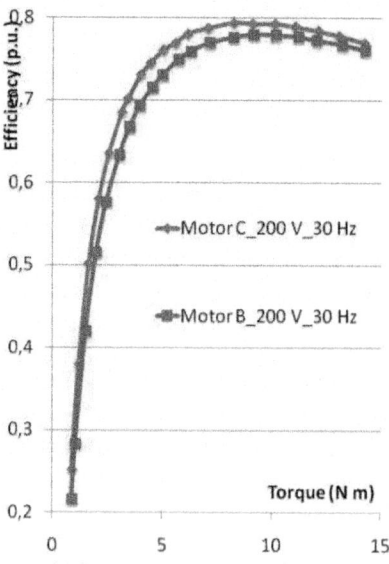

Fig. 12. Frequency & manufacturing technology impact

Figure 14: Frequency & manufacturing technology impact

ANNEX

Annex 1: NEMA MG-1 Table 12-11 Full-Load Efficiencies of Energy Efficient Motors (EPAct)

HP	ODP				TEFC			
	2 Pole	4 Pole	6 Pole	8 Pole	2 Pole	4 Pole	6 Pole	8 Pole
1.0	0.0	82.5	80.0	74.0	75.5	82.5	80	74
1.5	82.5	84.0	84.0	75.5	82.5	84	85.5	77
2.0	84.0	84.0	85.5	85.5	84	84	86.5	82.5
3.0	84.0	86.5	86.5	86.5	85.5	87.5	87.5	84
5.0	85.5	87.5	87.5	87.5	87.5	87.5	87.5	85.5
7.5	87.5	88.5	88.5	88.5	88.5	89.5	89.5	85.5
10.0	88.5	89.5	90.2	89.5	89.5	89.5	89.5	88.5
15.0	89.5	91.0	90.2	89.5	90.2	91	90.2	88.5
20.0	90.2	91.0	91.0	90.2	90.2	91	90.2	89.5
25.0	91.0	91.7	91.7	90.2	91	92.4	91.7	89.5
30.0	91.0	92.4	92.4	91.0	91	92.4	91.7	91
40.0	91.7	93.0	93.0	91.0	91.7	93	93	91
50.0	92.4	93.0	93.0	91.7	92.4	93	93	91.7
60.0	93.0	93.6	93.6	92.4	93	93.6	93.6	91.7
75.0	93.0	94.1	93.6	93.6	93	94.1	93.6	93
100.0	93.0	94.1	94.1	93.6	93.6	94.5	94.1	93
125.0	93.6	94.5	94.1	93.6	94.5	94.5	94.1	93.6
150.0	93.6	95.0	94.5	93.6	94.5	95	95	93.6
200.0	94.5	95.0	94.5	93.6	95	95	95	94.1
250.0	94.5	95.4	95.4	94.5	95.4	95	95	94.5
300.0	95.0	95.4	95.4	94.5	95.4	95.4	95	0
350.0	95.0	95.4	95.4	94.5	95.4	95.4	95	0
400.0	95.4	95.4	0.0	0.0	95.4	95.4	0	0
450.0	95.8	95.8	0.0	0.0	95.4	95.4	0	0
500.0	95.8	95.8	0.0	0.0	95.4	95.8	0	0

Annex 2: NEMA MG-1 Table 12-12 Full-Load Efficiencies for 60 Hz NEMA Premium Efficient Electric Motors Rated 600 Volts or less

HP	ODP			TEFC		
	2 Pole	4 Pole	6 Pole	2 Pole	4 Pole	6 Pole
1	77	85.5	82.5	77	85.5	82.5
1.5	84	86.5	86.5	84	86.5	87.5
2	85.5	86.5	87.5	85.5	86.5	88.5
3	85.5	89.5	88.5	86.5	89.5	89.5
5	86.5	89.5	89.5	88.5	89.5	89.5
7.5	88.5	91	90.2	89.5	91.7	91
10	89.5	91.7	91.7	90.2	91.7	91
15	90.2	93	91.7	91	92.4	91.7
20	91	93	92.4	91	93	91.7
25	91.7	93.6	93	91.7	93.6	93
30	91.7	94.1	93.6	91.7	93.6	93
40	92.4	94.1	94.1	92.4	94.1	94.1
50	93	94.5	94.1	93	94.5	94.1
60	93.6	95	94.5	93.6	95	94.5
75	93.6	95	94.5	93.6	95.4	94.5
100	93.6	95.4	95	94.1	95.4	95
125	94.1	95.4	95	95	95.4	95
150	94.1	95.8	95.4	95	95.8	95.8
200	95	95.8	95.4	95.4	96.2	95.8
250	95	95.8	95.4	95.8	96.2	95.8
300	95.4	95.8	95.4	95.8	96.2	95.8
350	95.4	95.8	95.4	95.8	96.2	95.8
400	95.8	95.8	95.8	95.8	96.2	95.8
450	95.8	96.2	96.2	95.8	96.2	95.8
500	95.8	96.2	96.2	95.8	96.2	95.8

Annex 3: CEI efficiencies

Table 3: Standard Efficiency IE1 50Hz

Rating (kW)	Rating (hp)	Poles		
		2	4	6
0.75	1	72.1	72.1	70.0
1.1	1.5	75.0	75.0	72.9
1.5	2	77.2	77.2	75.2
2.2	3	79.7	79.7	77.7
3	4	81.5	81.5	79.7
4	5.5	83.1	83.1	81.4
5.5	7.5	84.7	84.7	83.1
7.5	10	86.0	86.0	84.7
11	15	87.6	87.6	86.4
15	20	88.7	88.7	87.7
18.5	25	89.3	89.3	88.6
22	30	89.9	89.9	89.2
30	40	90.7	90.7	90.2
37	50	91.2	91.2	90.8
45	60	91.7	91.7	91.4
55	75	92.1	92.1	91.9
75	75	92.1	92.1	91.9
90	125	92.7	92.7	92.6
110	150	93.0	93.0	92.9
132	175	93.3	93.3	93.3
160	215	93.5	93.5	93.5
200	270	93.8	93.8	93.8
260	350	94.0	94.0	94.0
300	400	94.0	94.0	94.0
335	450	94.0	94.0	94.0
375	500	94.0	94.0	94.0

Table 4: Standard Efficiency IE1 60Hz

Rating (kW)	Rating (hp)	Poles		
		2	4	6
0.75	1	77.0	78.0	73.0
1.1	1.5	78.5	79.0	75.0
1.5	2	81.0	81.5	77.0
2.2	3	81.5	83.0	78.5
3	4	84.5	85.0	83.5
4	5.5	86.0	87.0	85.0
5.5	7.5	87.5	87.5	86.0
7.5	10	87.5	88.5	89.0
11	15	88.5	89.5	89.5
15	20	89.5	90.5	90.2
18.5	25	89.5	91.0	91.0
22	30	90.2	91.7	91.7
30	40	91.5	92.4	91.7
37	50	91.7	93.0	91.7
45	60	92.4	93.0	92.1
55	75	93.0	93.2	93.0
75	75	93.0	93.2	93.0
90	125	93.0	93.5	94.1
110	150	94.1	94.5	94.1
150	200	94.1	94.5	94.1
185	250	94.1	94.5	94.1
200	270	94.1	94.5	94.1
260	350	94.1	94.5	94.1
300	400	94.1	94.5	94.1
335	450	94.1	94.5	94.1
375	500	94.1	94.5	94.1

Table 5: High Efficiency IE2 50Hz

Rating (kW)	Rating (hp)	Poles		
		2	4	6
0.75	1	77.4	79.6	75.9
1.1	1.5	79.6	81.4	78.1
1.5	2	81.3	82.8	79.8
2.2	3	83.2	84.3	81.8
3	4	84.6	85.5	83.3
4	5.5	85.8	86.6	84.6
5.5	7.5	87.0	87.7	86.0
7.5	10	88.1	88.7	87.2
11	15	89.4	898.0	88.7
15	20	90.3	90.6	89.7
18.5	25	90.9	91.2	90.4
22	30	91.3	91.6	90.9
30	40	92.0	92.3	91.7
37	50	92.5	92.7	92.2
45	60	92.9	93.1	92.7
55	75	93.2	93.5	93.1
75	75	93.8	94.0	93.7
90	125	94.1	94.2	94.0
110	150	94.3	94.5	94.3
132	175	94.6	94.7	94.6
160	215	94.8	94.9	94.8
200	270	95.0	95.1	95.0
260	350	95.0	95.1	95.0
300	400	95.0	95.1	95.0
335	450	95.0	95.1	95.0
375	500	95.0	95.1	95.0

Table 6: High Efficiency IE2 60Hz (EPACT)

Rating (kW)	Rating (hp)	Poles		
		2	4	6
0.75	1	75.5	82.5	80.0
1.1	1.5	82.5	84.0	85.5
1.5	2	84.0	84.0	86.5
2.2	3	85.5	87.5	87.5
3.7	5	87.5	87.5	87.5
5.5	7.5	88.5	89.5	89.5
7.5	10	89.5	89.5	89.5
11	15	90.2	91.0	90.2
15	20	90.2	91.0	90.2
18.5	25	91.0	92.4	91.7
22	30	91.0	92.4	91.7
30	40	91.7	93.0	93.0
37	50	92.4	93.0	93.0
45	60	93.0	93.6	93.6
55	75	93.0	94.1	93.6
75	75	93.6	94.5	94.1
90	125	94.5	94.5	94.1
110	150	94.5	95.0	95.0
150	200	95.0	95.0	95.0
185	250	95.4	95.4	95.0
200	270	95.4	95.4	95.0
260	350	95.4	95.4	95.0
300	400	95.4	95.4	95.0
335	450	95.4	95.4	95.0
375	500	95.4	95.4	95.0

Table 7: Premium Efficiency IE3 50Hz

Rating (kW)	Rating (hp)	Poles 2	Poles 4	Poles 6
0.75	1	80.7	82.5	78.9
1.1	1.5	82.7	84.1	81.0
1.5	2	84.2	85.3	82.5
2.2	3	85.9	86.7	84.3
3	4	87.1	87.7	85.6
4	5.5	88.7	88.6	86.8
5.5	7.5	89.2	89.6	88.0
7.5	10	90.1	90.4	89.1
11	15	91.2	91.4	90.3
15	20	91.9	92.1	91.2
18.5	25	92.4	92.6	91.7
22	30	92.7	93.0	92.2
30	40	93.3	93.6	92.9
37	50	93.7	93.9	93.3
45	60	94.0	94.2	93.7
55	75	94.3	94.6	94.1
75	75	94.7	95.0	94.6
90	125	95.0	95.2	94.9
110	150	95.2	95.4	95.1
132	175	95.4	95.6	95.4
160	215	95.6	95.8	95.6
200	270	95.8	96.0	95.8
260	350	95.8	96.0	95.8
300	400	95.8	96.0	95.8
335	450	95.8	96.0	95.8
375	500	95.8	96.0	95.8

Table 8: Premium Efficiency IE3 60Hz (NEMA Premium)

Rating (kW)	Rating (hp)	Poles 2	Poles 4	Poles 6
0,75	1	77,0	85,5	82,5
1,1	1,5	84,0	86,5	87,5
1,5	2	85,5	86,5	88,5
2,2	3	86,5	89,5	89,5
3,7	5	88,5	89,5	89,5
5,5	7,5	89,5	91,7	91,0
7,5	10	90,2	91,7	91,0
11	15	91,0	92,4	91,7
15	20	91,0	93,0	91,7
18,5	25	91,7	93,6	93,0
22	30	91,7	93,6	93,0
30	40	92,4	94,1	94,1
37	50	93,0	94,5	94,1
45	60	93,6	95,0	94,5
55	75	93,6	95,4	94,5
75	75	94,1	95,4	95,0
90	125	95,0	95,4	95,0
110	150	95,0	95,8	95,8
150	200	95,4	96,2	95,8
185	250	95,8	96,2	95,8
200	270	95,8	96,2	95,8
260	350	95,8	96,2	95,8
300	400	95,8	96,2	95,8
335	450	95,8	96,2	95,8
375	500	95,8	96,2	95,8

Annex 4: Acronyms
- ASD Adjustable speed drives
- ACEEE American Council for and Energy Efficient Economy
- CEE Consortium for Energy Efficiency
- DOE Department of Energy
- EISA Energy Independence & Security Act
- EPAct Energy Policy act
- EU European Union
- GHG Greenhouse gas emissions
- IPCC International Panel on climate change
- OEM Original equipment manufacturer
- MG 1 Motor Generator
- MEPS Minimum energy performance standard MEPS
- MDM Motor Decisions Matter (MDM
- NEMA National Electrical Manufacturers Association

REFERENCES

1. Aoulkadi, M., Binder, A.: "Evaluation of different measurement methods to determine stray load losses in induction machines,". IEEE Trans. On Industrial Electronics, vol. 2 No 1, 2008
2. Benhaddadi M., Olivier G.: (2010a) "La promotion de l'économie d'énergie électrique passe par son juste prix,". 79e Congrès de l'AFAS, Montréal, Canada, 2010
3. Benhaddadi M., Olivier G., and Yelle J.: (2010b) "Premium efficiency motors effectiveness,". IEEE International symposium on power electronics, electrical drives, automation and motion SPEEDAM 2010, Pisa, Italy, 2010
4. Benhaddadi M., Olivier G., Labrosse D., Tétrault P.: (2009a) "Premium efficiency motors and energy saving potential,". IEEE International electric machines and drives conference, IEEE_IEMDC, Miami, USA, 2009
5. Benhaddadi M., Olivier G.: (2009b) "L'économie d'énergie : une affaire de législation,". Communication présentée au 78e Congrès de l'ACFAS, Ottawa, Canada, 2009
6. Benhaddadi M., Olivier G.: (2008a) "Barriers and incentive policies to high-efficiency motors and drives market penetration,". IEEE

International symposium on power electronics, electrical drives, automation and motion SPEEDAM 2008, Ischia, Italy,
7. Benhaddadi M., Olivier G.: (2008b) "Le génie électrique à la rescousse des économies d'énergie dans l'industrie,". Communication présentée au 76ᵉ Congrès de l'ACFAS, Québec, Canada, 2008
8. Benhaddadi M., Olivier G.: (2008c) "Dilemmes énergétiques,". Presses de l'Université du Québec, Québec, 2008, 216p.
9. Benhaddadi, M., Olivier G.: (2007) "Energy savings by means of generalization adjustable speed drive Utilization,". IEEE Canadian Conference on Electrical and Computer Engineering, Vancouver, 2007.
10. Benhaddadi M., Olivier G.: (2004) "Including Kyoto in electrical engineering curriculum,". IEEE Canadian Conference on Electrical and Computer Engineering, Niagara Falls, 2004
11. Boglietti, A. Cavagnino, A. Lazzari, M. Pastorelli, , M.: (2004) "International standards for the induction motor efficiency evaluation: a critical analysis o A. f the stray-load loss determination,". IEEE Trans. On Industry Appl., vol.40, No 5, 2004
12. Bonnett, A.H., Yung, C.: (2008) "Increased efficiency versus increased reliability: A comparison of pre-EPAct, EPAct, and premium-efficiency motors,". IEEE Industry application Magazine, vol.2, 2008
13. BP (2010) British Petroleum statistical review of world energy (2010), June 2010
14. CEE (2007) Energy-efficiency incentive programs: Premium-efficiency motors and adjustable speed drives in the US and Canada, (2007) prepared by consortium for energy efficiency, may 2007, www.cee.org
15. DOE (2010) Energy Information Administration, Official energy statistics from the US government, DOE/EIA http://www.eia.doe.gov
16. Elmeida, A.I. Ferreira, F.FJ. Busch, J.F. Angers P. (2002) "Comparative analysis of IEEE-112 B and IEC 34-2 efficiency testing standards using stray load losses in low-voltage three phase cage induction motors," IEEE Industry applications, March-April 2002, pp. 608-614
17. Energy Policy Act EPACT and motor testing understanding, IEEE-112 Method B, Evans, B.D., Crissman, J., Gobert, G.: (2008) "Test results for energy savings,". IEEE Industry application Magazine, vol.2, 2008
18. Finlay, W.R. Veerkamp B., Gehring D., and Hanna: P. (2009)"Improving motor efficiency levels globally,". IEEE Industry application Magazine, vol.15, 2009
19. IEA (2010) International Energy Agency, OECD/IEA http://oecd.org./

20. IEA (2009) The experience with energy efficiency policies and programs in IEA countries: learning from the critics, (2009) International energy agency information paper, august 2009
21. IPCC (2007) Contribution of working group I to the fourth assessment report of the intergovernmental panel on climate change, summary for policymakers, http://www.ipcc.ch
22. Nagorny, A., Wallace, A., Von Jouanne, A.: "Stray load loss efficiency connections,". IEEE Industry application Magazine, vol.10, issue 3, may-june 2004
23. NRC (2004) Energy-efficient motor systems assessment guide, Canadian industry program for energy conservation (2004), CIPEC, Natural Resources Canada, 2004
24. Parasiliti, F., Villani, M., Paris, C., Walti, O. Songini, G. Novello, A. Rossi T.: (2004) "Three-phase induction motor efficiency improvements with die-cast cooper rotor cage and premium steel,". IEEE International symposium on power electronics, electrical drives, automation and motion SPPEDAM 2004, Capri, Italy, 2004
25. RNC (2004) Ressources Naturelles Canada: (2004) "Guide d'évaluation du rendement des systèmes moteurs éconergétiques," RNC, 2004
26. Renier, B. Hameyer, K., Belmans, R.: (1999) "Comparison standards for determining efficiency of three phase induction motors,". IEEE Trans. On Energy Conversion, vol.14, No 3, 1999
27. Rooks, J.A., Wallace: (2004) "Energy efficiency of VSDs," IEEE Industry applications Magazine, vol.10, issue 3, 2004
28. UE (2003) Commission Européenne, Direction générale énergie et transport: (2003) "Motor challenge programme,". UE, Bruxelles, 2003

Chapter 2

A NEW TYPE OF CAPACITIVE MACHINE

Arie Shenkman, Saad Tapuchi, and Dmitry Baimel

Electrical and Electronics Engineering Department, Shamoon College of Engineering, Beer Sheva, Israel

ABSTRACT

The paper proposes a new type of the synchronous capacitive machine operated on a principle of the electric field effect. The proposed machine has smaller size and lighter weight than the standard electromagnetic synchronous machines with the same rated parameters. Another important advantage is a simple structure of the machine, which simplifies the production process and reduces the costs of the motor. The paper also presents extensive simulation results of the proposed capacitive machine. The simulation results show that the proposed machine is able to reach the same power output as the electromagnetic machines.

INTRODUCTION

Today, most of the electrical machines operate on the principle of magnetic field effect at which the electric energy is converted into mechanical energy and vice versa. The magnetic field is produced by the current flowing in the stator and rotor coils.

On the other hand, it is also possible to construct the electrical machines that operate on the electric field principle, where the electrical energy is stored in an electric field [1] . It is known that the electric and the magnetic fields are always linked to each other and it is impossible to separate them. However, the electric field in the electromagnetic machines is much weaker than the magnetic field, and, practically, is negligible.

The electric machines operated on the electric field principle would have mechanical structure based on capacitors [2] - [4] while the mechanical structure of the electromagnetic machines is based on the inductors. While the development of the electromagnetic machines has reached the most advanced state, the electric field machines (capacitor machines) remained somehow ignored. The reason is low electric permittivity-ε of most dielectric materials. However, recently, new dielectric materials have been developed. These materials possess relatively high electric permittivity of up to $\varepsilon_r = 10^4$. Therefore, the capacitor motors acquire an interest again.

The greatest difficulty in the development and production of capacitive machines is preventing friction between the stator and rotor plates which occurs when the dielectric materials are solid. A common method to prevent the friction is to ensure an air gap. However, in the capacitive machine this method is not suitable because even small air gap diminishes the advantages of using dielectric material with high dielectric permittivity ε. One possible solution could be to use liquid electric materials with low density and high values of ε. Another possible solution might be changing the air gap to plasma, which can be accomplished by using high voltage and a narrow air gap.

The main factor determining the efficiency of electric power mechanisms for energy conversion is the density of the electric or magnetic energy stored in the electromagnetic or electric fields. The electromagnetic and electric field machines could be compared from the energy point of view [5] [6]. The energy density of the magnetic field in electromagnetic machines could be calculated by:

$$w_m = \frac{B^2}{2\mu_0 \mu_r} \quad (1)$$

where B is the magnetic flux density of the magnetic field, μ_0 is the magnetic permeability constant $(\mu_0 = 4\pi \times 10^{-7}\ \text{H/m})$ and μ_r is the magnetic permeability of the air (in magnetic machines the process of energy conversion occurs within the air gap between the rotor and the stator so $\mu_r = 1$). By using the ferromagnetic materials (particularly electric steel), the magnetic field of the electromagnetic machines can have magnetic flux density of $B = 1-1.6\ \text{Wb/m}^2$. The density of the magnetic energy would be (for $B = 1.4\ \text{Wb/m}^2$):

$$w_m = \frac{1.4^2}{2 \times 4\pi \times 10^7} = 800000\ \text{W/m}^3 \quad (2)$$

The energy density of the electric field in the electric machines could be calculated by:

$$w_e = \frac{1}{2} \varepsilon_r \varepsilon_0 E^2 \quad (3)$$

where ε_0 is the electric permittivity constant $(\varepsilon_0 = 8.86 \times 10^{-12}\ \text{F/m})$ and E is electric flux density of the electrostatic field. Today's technology allows usage of very strong electrostatic fields, but because of low electric strength of most insulating materials, the electrostatic field is practically limited to 2 - 20 MV/m. Therefore, in the machine whose operation is based on the electric field principle and in which the process of energy conversion also takes place in the air $(\varepsilon_r = 1)$, the energy density of the electric field, for $E = 10^7$ V/m, could be determined by Equation (4):

$$w_e = \frac{8.86 \times 10^{-12} \times (10^7)}{2} = 450 \text{ W/m}^3 \tag{4}$$

Lately, new dielectric materials have been developed, such as ceramics based on a paratitanite bar, which has a relatively high electric permittivity of up to $\varepsilon_r = 10^4$ [F/m]. In electric machines, where the energy conversion process takes place in the new dielectric materials and not in the air, the density of the electric energy could be significantly increased.

For electric permittivity of $\varepsilon_r = 10^4$, the obtained density of the electric energy, for $E = 10^7$ V/m, would be:

$$w_e = \frac{10^4 \times 8.86 \times 10^{-12} \times (10^7)}{2} = 450 \times 10^4 \text{ W/m}^3 \tag{5}$$

Such a result is even greater than the result obtained with the electromagnetic machines, a fact that encourages development of new electric machines based on the electric field effect.

The electric field machines have several advantages upon the electromagnetic machines. The main advantage is that the electric field machines are much smaller in size and lighter in weight than the electromagnetic machines with the same output parameters. This could be explained by the fact that magnetic force lines of any electromagnetic machine are formed in a closed loop. The magnetic force lines are formed by the magnetic circuit, built from a very heavy material like electric steel. The magnetic circuit is used for the power transfer, while the energy conversion process takes place in the air-gap itself (between the stator and the rotor). The air-gap volume is relatively a small part of the volume of the whole machine.

In the electric field machines, the power lines of the electric field start and end in electric charges, i.e. the rotor and stator plates. The energy conversion process takes place in the gap between the stator and rotor, which takes up a sizable part of the volume of the whole machine. Assuming that the specific energy is equal for both electric field and electromagnetic machines, the machines based on the electric field effect will be significantly smaller. And since the dielectric materials (e.g. ceramics and plastics) are much lighter than the ferromagnetic materials, the electric machines based on the electric field effect will be also lighter. As a result of these two factors, the production of the machines based on the electric field effect will be cheaper.

Just like magnetic machines, the capacitive machines can be of two kinds, in dependence of the current/vol- tage used for their operation: direct current (DC) machines with commutator or alternative current (AC) machines: synchronous or asynchronous.

Section II shows how the sectored capacitor could be used as a practical solution for the DC capacitive machines. Section III presents the proposed AC capacitive machine based on the sinusoidal capacitor. Section IV presents the simulation results of the proposed capacitive machine. Section V presents the conclusions of the paper.

THE SECTORED CAPACITOR AS A PRACTICAL SOLUTION FOR THE DC CAPACITIVE MACHINES

The simplest DC capacitive machine can be constructed from a round capacitor having two or more sectors, as shown in Figure 1. In this machine, two positively charged rotating sectors form the rotor and two negatively charged static sectors form the stator. In the position shown in Figure 1, the attractive force between the stator and rotor sectors establishes a moment, which rotates the rotor on a quarter of a circle until the sectors coincide. At the moment of a coincidence of the sectors, the applied voltage polarity is changed between the positively and negatively charged sectors. In other words, this machine should be constructed with a commutate or (like regular DC machines). If the number of plates will be increased, the machine operation will be smoother. In order to ensure smooth operation, at least three couples of plates should be used.

The torques of the capacitive machine could be calculated by differentiation of the energy stored in the electric field of the capacitor [6] . In the case of angular movement, the differentiation is performed by an angle α:

$$M(\alpha) = \frac{d}{d\alpha}\left(\frac{C(\alpha)V^2}{2}\right) \qquad (6)$$

where $C(\alpha)$ is varying capacitance of the capacitor plates as function of the rotor angle α and V is the voltage applied to the capacitor plates (this formula is adequate when the capacitor is connected to a power source). With the movement of the rotor, the capacitance between the rotor and the stator varies proportionally to the surface of coincidence between sectors. The surface of coincidence between the sectors is proportional to the rotor angle α and could be calculated by:

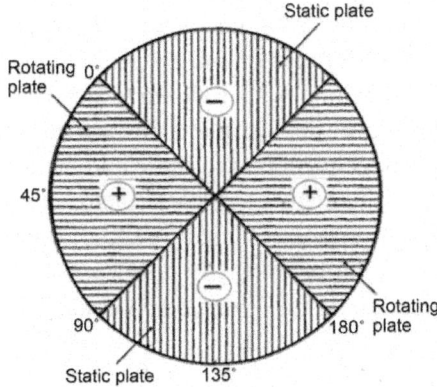

Figure 1: DC capacitive machine constructed from round capacitor with four sectors.

$$s = \rho^2 \frac{\alpha}{2} \quad (7)$$

where ρ is the radius of the plates. The capacitance of the round capacitor with four sectors (two couples of plates) could be calculated by:

$$C(\alpha) = 2\frac{\varepsilon\varepsilon_0 \rho^2 \alpha}{d} \quad (8)$$

where d is the distance between the plates.

The axial torque of the capacitive machine with three couples of plates could be calculated by substituting Equation (8) in Equation (6):

$$M(\alpha) = \frac{\varepsilon\varepsilon_0 \rho^2 V^2}{d} \quad (9)$$

Equation (9) shows that the moment of this capacitive machine is constant and independent of the rotor angle-α. The speed of capacitive motors is a function of the supplied power and the number of revolutions per minute could be calculated by $= 9.55 \times (P/M)$. For example, for the 130 W motor with insulating material which has dielectric permittivity of $\varepsilon = 10^4$, two couples of plates, the distance between the plates $d = 2$ mm, the radius of the plates $\rho = 10$ cm and applied voltage $V = 1$ kV, the moment will be $M = 0.44$ Nm. The motor speed will be $n = 955$ rpm.

THE PROPOSED AC CAPACITIVE MACHINE BASED ON A SINUSOIDAL CAPACITOR

This paper proposes novel synchronous machine operating on the principle of the electric field effect. The proposed machine is constructed from a variable capacitor having static and rotating plates. The capacitance be haves as a sinusoidal function and depends on the plates movement angle. The proposed sinusoidal capacitor machine with rotating plates (rotor) above static plates (stator) is shown in Figure 2. The plates are constructed in such a way that their radius-vector is dependent on the angle and their overlapping surface is a sinusoidal function of the angle. The stator plates are constructed as standard sectors. This form allows more efficient usage of the plate's surface and as a result, increases the maximum capacity of the capacitor. The variable surface of the plates is a function of the α angle and could be calculated by:

$$S(\alpha) = \frac{1}{2}\int_0^\alpha \rho^2(\alpha)d\alpha \qquad (10)$$

On the other hand, the variable surface is sinusoidal and it could be also expressed as:

$$S(\alpha) = S_m \frac{1-\cos(2\alpha)}{2} \qquad (11)$$

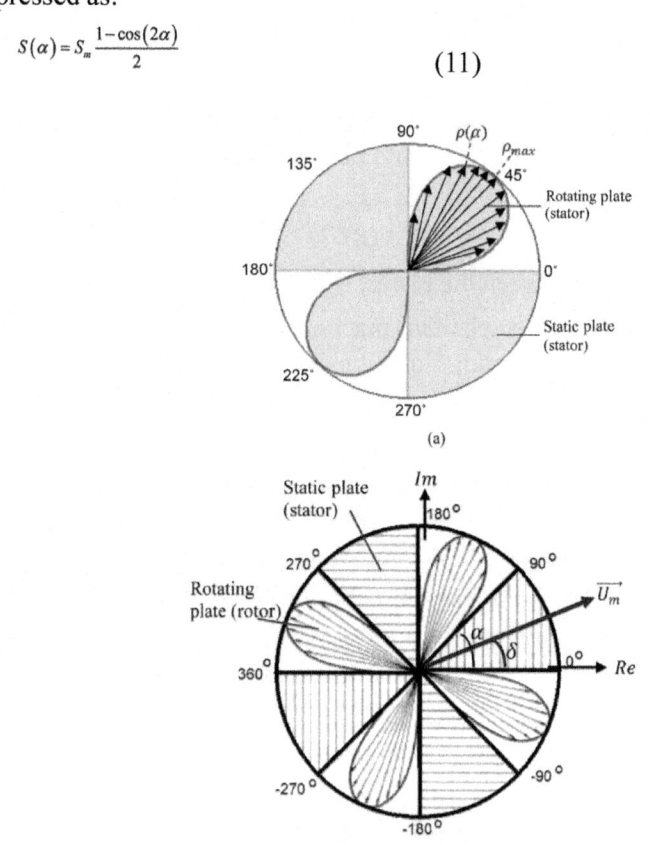

Figure 2: (a) The proposed capacitive motor with two rotating plates (rotor) and two

static plates (stator); (b) The proposed capacitive motor with four rotating plates (rotor) and four static plates (stator).

where S_m is the maximum overlap surface of the capacitor (when the plates of the stator and the rotor are fully overlapped one above the other). The variable surface versus the rotating angle α is shown in Figure 3.

The radius of the capacitor plates $\rho(\alpha)$ could be calculated from Equations (10) and (11):

$$\rho(\alpha) = \sqrt{2 S_m \sin(\alpha)} \qquad (12)$$

By using Equation (12), the diameter of the capacitor plates could be obtained:

$$D = 2\rho_m = 2\sqrt{2 S_m} \qquad (13)$$

where ρ_m is the maximal radius of the rotating plates.

When the rotor plates rotates above the stator plates, the capacity of the machine varies according to:

$$C(\alpha) = C_m \frac{1 - \cos(2\alpha)}{2} = \varepsilon \varepsilon_0 k \frac{S_m}{d} \frac{1 - \cos(2\alpha)}{2} \qquad (14)$$

where d is the distance between the static and the rotating plates, ε is the relative electric permittivity of the insulating material between the plates, k is the number of pairs of the rotor/stator plates and C_m is the maximal capacitance of the machine with k couple of plates.

The torque of the capacitive machine could be calculated by the differentiation of the energy stored in the electric field of the capacitor. If the capacitor is supplied by the voltage-u, the axial torque of the capacitive machine could be determined by substituting Equation (14) into Equation (6):

$$M(\alpha) = \frac{C_m u^2}{2} \sin 2\alpha \qquad (15)$$

As it seen from the Equation (15), if the proposed capacitive machine was supplied by a DC voltage source, the average torque would be zero. In order to obtain efficient average torque (different than zero), the applied voltage should be an alternating voltage $(u = U_m \sin(\omega t))$ and the angular speed of the rotor should be equal to the angular frequency of the applied voltage.

The angular velocity could be defined as $\omega = \frac{d\alpha}{dt}$. Therefore, the rotor angle could be obtained by an integration of the angular velocity:

$$\alpha = \omega t + \frac{\delta}{2} \qquad (16)$$

where δ is the angle between the starting line $(\alpha = 0°)$ of the rotor plate and the phase angle of the alternating voltage (see Figure 3(b)). The meaning of the angle δ could be also explained by using a complex plane having "Real" and "Imaginary" axes. The alternating voltage applied to the sinusoidal capacitor could be represented by the rotating vector U_m, which rotates with the angular velocity ω on the complex plain. The capacitor rotating plates (rotor) are referred to the zero degree line. Therefore, the δ angle is the angle between voltage vector U_m and the rotor axis.

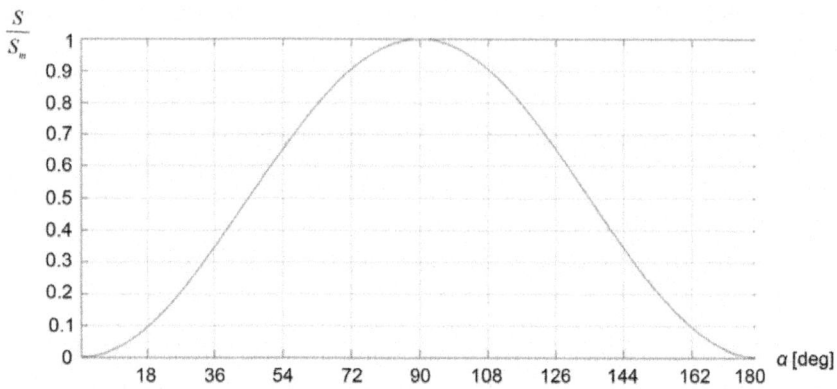

Figure 3: The variable surface of the capacitor versus the angle α.

The angular velocity of the rotor could be obtained by:
$$\omega_r = \omega/k \qquad (17)$$
where k is the number of pairs of the rotor/stator plates. The speed of the machine in revolutions per minute (rpm) could be calculated according to the following formula:
$$n = \frac{60 f}{k} \qquad (18)$$
where f is the frequency of the applied AC voltage.

The torque of the proposed machine as a function of time could be calculated by substituting the Equation (16) into the Equation (15):
$$M(t) = C_m U_m^2 \sin(\omega t)^2 \sin(2(\omega t + \delta/2)) \qquad (19)$$
$$M(t) = \frac{C_m U_m^2}{2}(1 - \cos(2\omega t))\sin(2\omega t + \delta) \qquad (20)$$

$$M(t) = \frac{C_m U_m^2}{2}\left[\sin(2\omega t + \delta) - \frac{\sin\delta}{2} - \frac{\sin(4\omega t + \delta)}{2}\right] \quad (21)$$

The average torque, i.e., the effective torque, would be:

$$M_{avg} = \frac{1}{2\pi}\int_0^{2\pi} M(t)dt = -\frac{C_m U_m^2}{2}\sin\delta \quad (22)$$

The meaning of the minus sign in the formula of torque is that the torque resists motion, i.e. that machine operates in a generator mode and the conversion is from mechanical energy into electric energy. In the generator mode, the rotor precedes the alternating voltage (positive δ angle). In the motor mode, where the δ angle is negative, the rotor lags behind the alternating voltage. The generator and the motor operation modes are shown in the torque diagram, Figure 4.

The machine would develop maximal torque when the angle δ would be equal to $\pi/2$:

$$M_{max} = \frac{U^2 C_m}{2} \quad (23)$$

The obtained diagram is similar to the torque diagram of a synchronous machine. It can be seen from the presented analysis, that a rotating field is not needed for the capacitive machine operation as for the three-phase electromagnetic machines, and that a capacitor machine can also operate as a single-phase machine. There are no differences between the analysis of single-phase capacitor motor and the three-phase one.

SIMULATION OF THE PROPOSED AC CAPACITIVE MACHINE

The proposed AC capacitive machine was simulated by using the Matlab program. The parameters of the simulated machine are: 4 couples of plates $(k=4)$, the maximal overlap surface of the rotor and stator plates is $S_m = 50 \text{ cm}^2$, electric permittivity of the dielectric material is $\varepsilon = 10^4$, the distance between the plates is $d = 3 \text{ mm}$.

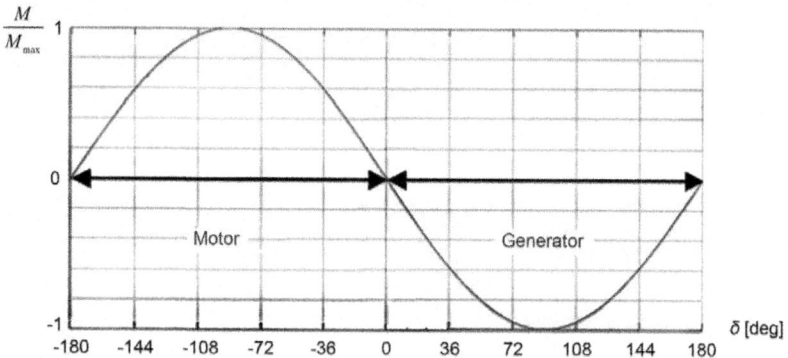

Figure 4: The torque diagram of the proposed capacitive machine.

The applied AC voltage is $U_m = 6\,kV$ and the frequency is $f = 50\,Hz$. The simulation results for $\delta = 30°$ (generator mode) are shown in Figure 5. It can be seen that the capacitance varies in a sinusoidal shape: $C(\alpha) = 0$ for $\alpha = 0, \pi, 2\pi$; $C(\alpha) = 6\,uF$ for $\alpha = 0, 0.5\pi, 1.5\pi$. The obtained average torque is negative -54 Nm (see Figure 5(a)). Therefore, the machine operates in the generator mode. The spectrum of the instantaneous torque has three harmonics: DC, 100 Hz and 200 Hz. This result is logical because $M(t)$ consists of $-\frac{C_m U_m^2 \sin\delta}{2\;\;\;2}$ which is represented by DC harmonic, $\frac{C_m U_m^2}{2}[\sin(2\omega t + \delta)]$ which is represented by 100 Hz harmonic and $-\frac{C_m U_m^2 \sin(4\omega t + \delta)}{2\;\;\;2}$ which is represented by 200 Hz harmonic.

The simulation results for $\delta = 0°$ are shown in Figure 6. In this case, the machine operates on the boundary of generator and motor modes. The obtained average torque is zero. The spectrum of the instantaneous torque has 100 Hz and 200 Hz harmonics. The DC harmonic is zero because $\frac{C_m U_m^2 \sin\delta}{2\;\;\;2} = 0$.

The simulation results for $\delta = -30°$ are shown in Figure 7. In this case, the machine operates in the motor-mode. The obtained average torque is 54 Nm.

The Matlab simulation results could be validated by numerical calculations. The maximal capacitance could be calculated by Equation (14):

$$C_m = \varepsilon\varepsilon_0 k\frac{S_m}{d} = 10^4 \times 8.86 \times 10^{-12} \times 4\frac{50 \times 10^{-4}}{3 \times 10^{-4}} = 6\,uF$$

A New Type of Capacitive Machine   41

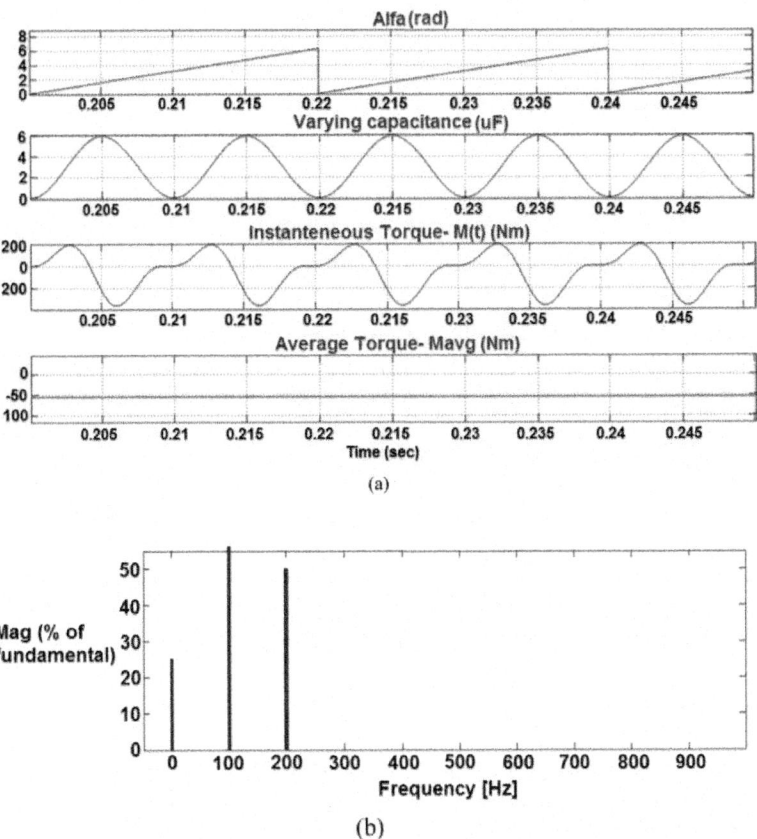

Figure 5. The angle $\delta = 30°$ (generator mode). (a) The simulated rotating rotor angle α (varies from zero to 2π), varying capacitance $C(\alpha)$, instantaneous torque of the machine $M(t)$ and the average torque M_{avg}. (b) The spectrum of the instantaneous torque.

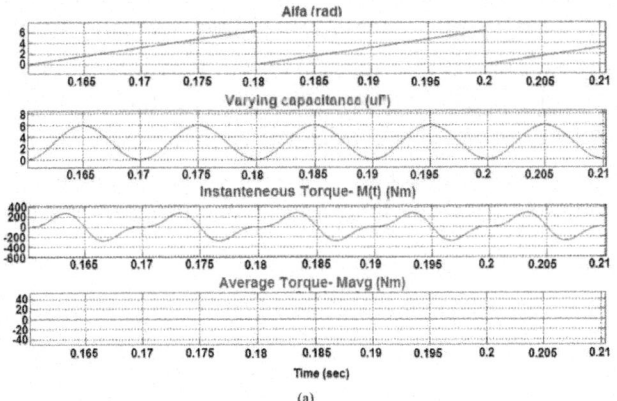

(a)

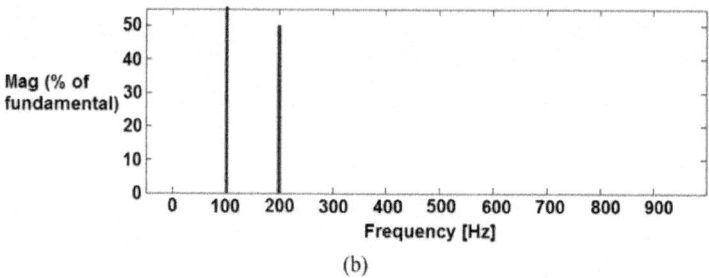

(b)

Figure 6: The angle $\delta = 0°$ (boundary between the generator and motor modes). (a) The simulated rotating rotor angle α (varies from zero to 2π), varying capacitance $C(\alpha)$, instantaneous torque of the machine $M(t)$ and the average torque M_{avg}; (b) The spectrum of the instantaneous torque.

For the case of $\delta = 30°$, the calculated average torque would be:

$$M_{avg} = -\frac{C_m U_m^2}{2}\sin\delta = -\frac{(6\times10^{-6})\times(6\times10^3)^2}{2}\times 0.5 = -54 \text{ Nm}$$

For the case of $\delta = -30°$, the calculated average torque would be:

$$M_{avg} = -\frac{C_m U_m^2}{2}\sin\delta = -\frac{(6\times10^{-6})\times(6\times10^3)^2(-0.5)}{2} = 54 \text{ Nm}$$

In this case, the motor output power would be:

$$P_{out} = M_{avg} \times \omega_r = M_{avg} \times \frac{\omega}{k} = \frac{54\times 314}{4} = 4239 \text{ W}$$

The calculations show that the presented simulation results are correct.

CONCLUSIONS

The novel capacitive synchronous machine is proposed. The machine is operating on the principle of the electric field. By using new materials with high electric permittivity ε, the proposed machine is feasible from the theoretical and the practical points of view. The machine design is based on the sinusoidal capacitor whose capacitance varies according to a sinusoidal function. The proposed machine is actually electrical synchronous machine with characteristics similar to those of the magnetic synchronous machines.

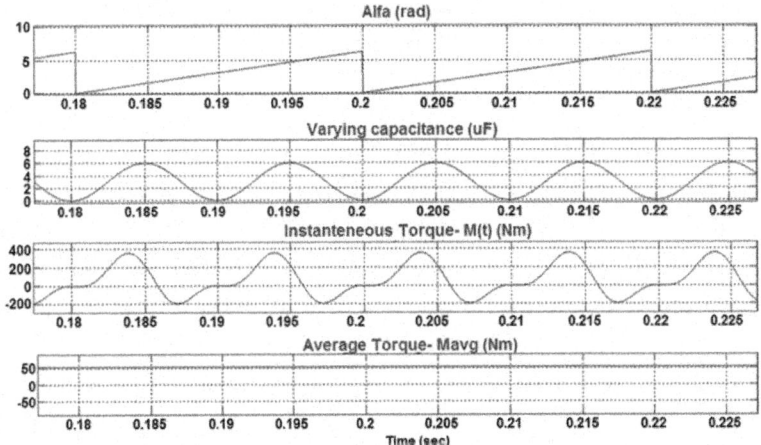

Figure 7: The angle $\delta = -30°$ (motor modes). The simulated rotating rotor angle α (varies from zero to 2π), varying capacitance $C(\alpha)$, instantaneous torque of the machine $M(t)$ and the average torque M_{avg}.

Extensive simulation results of the machine operated in the motor and the generator modes validate the theoretical analysis and show the practical feasibility of the proposed machine.

The proposed machine has a number of advantages upon the conventional electric machines based on the magnetic interaction. The main advantage is that the electric field machines are much smaller in size and lighter in weight than the electromagnetic machines with the same output parameters. Furthermore, the capacitor motor has very low inertial torque due to the low weight of the rotor. This fact allows rapid starting and stopping of the machine, as well as other variations during operation. This feature could be highly important in control systems. Another advantage is the simple structure of a single-phase capacitor motor.

REFERENCES

1. Moore, A.D. (1973) Electrostatics and Its Applications. John Wiley & Sons, 133-145.
2. Hojjat, Y., Dadkhah, M., Modabberifar, M. and Ghanbari, M. (2007) Electrostatic Rotation of Plexiglas Disc Supported on Air Bearing. 14th International Conference on Mechatronics and Machine Vision in Practice, Xiamen, 4-6 December 2007, 130-135.
3. Santana-Martín, F.J., Montoya, S.G.-A., Monzón-Verona, J.M. and Montiel-Nelson, J.A. (2008) Analysis and Modeling of an Electrostatic

Induction Micromotor. 18th International Conference on Electrical Machines, Vilamoura, 6-9 September 2008, 1-5.
4. Stranczl, M., Sarajlic, E., Fujita, H. and Yamahata, C. (2011) Modal Analysis and Modeling of Frictionless Electrostatic Rotary Stepper Micromotor. IEEE 24th International Conference on Micro Electro Mechanical Systems (MEMS), Cancun, 23-27 January 2011, 1257-1260.
5. Matsuzaki, K., Matsuo, T. and Mikuria, Y. (1994) Comparison of Electrostatic and Electromagnetic Motors Based on Fabrication and Performance Criteria. Proceedings of 5th International Symposium on Micro Machine and Human Science, Nagoya, 2-4 October 1994, 77-81.
6. Behjat, V. and Vahedi, A. (2004) Study Torque Ripple of the Side Drive Electrostatic Micromotor. 39th International Universities Power Engineering Conference, Bristol, 8 September 2004, 600-603.

Chapter 3

OPTIMAL DESIGN AND CONTROL OF A TORQUE MOTOR FOR MACHINE TOOLS

Yee-Pien YANG, Shih-Chin YANG, and Jieng-Jang LIU

Department of Mechanical Engineering, National Taiwan University, Taipei, Taiwan, China.

ABSTRACT

This paper presents a systematic approach of optimal design and control of a surface-mount, permanent-magnet synchronous torque motor for the next-generation machine tools. A step-by-step procedure of optimization integrates multiple performance objectives and constraints to help the designer make the best decision on the final motor geometry from both design and control perspectives. In the perspective of design, a torque motor with concentrated windings and similar numbers of slots and poles may achieve the desired performance after optimization for multiple objectives, leading to a sinusoidal flux density for a nearly ripple-free torque distribution. From the control perspective, an optimal current waveform with an ideal shift angle is determined for each phase by aligning the current excitation with the back electromotive force. Both design and control of the surface-mount, permanent magnet machine are verified by the finite element method, and a prototype is fabricated for performance validation.

INTRODUCTION

In many industrial applications, torque motors attract much attention by their appealing features of high torque density, high efficiency and low ripples. A particular type of permanent magnet (PM) motor uses advanced electronic commutation to replace traditional mechanical commutation. This simple structure makes it easier to fabricate at low cost in the direct or indirect-drive arrangement. The direct-drive arrangement eliminates transmission trains and gearboxes to operate at a lower speed with higher torque in various implementations of machine tools and transportation vehicles. The geometry of assembly of a surface-mount, permanent magnet (SPM) torque motor is illustrated in Figure 1.

To achieve its optimal performance, a torque motor needs a systematic methodology of design and control. From the design perspective, the motor configuration and geometry are determined in an optimal way by maximizing the output torque with minimal weight and least power consumption. First, the configuration of concentrated windings, instead of distributed windings, is determined because of its advantages of short end windings and simple structure. Similar numbers of slots and poles are also selected [1–3], so that the slot pitch is close to the pole pitch, not only to reduce the copper loss but also to produce the largest torque in the concentrated windings. The symmetric winding layout of similar numbers of slots and poles is very suitable for conventional three-phase drives [4]. Second, the motor geometries, such as the shape of the magnet and stator tooth, are crucial parameters for motor performance. Hsieh and Hsu [5] and Islam et al. [6] made several investigations on a motor with sinusoidal current excitations and found that the waveform of magnetic flux was very sensitive to the shape of the magnet. Recently, Yang et al. [7] invented a rim motor of sectional stators and arc magnets; its magnetic flux distribution in the air gap was close to a sinusoidal function, thereby producing a near-sinusoidal back electromotive force (EMF) with fewer harmonics.

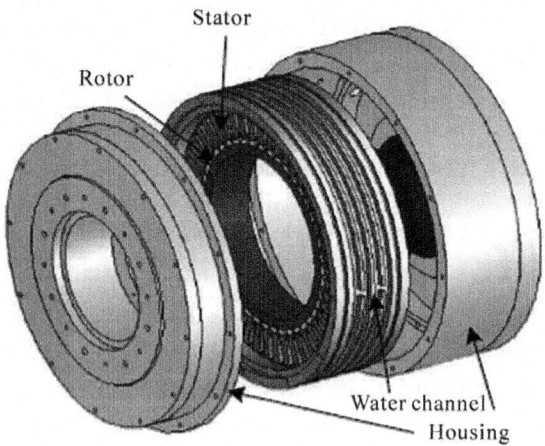

Figure 1: The SPM machine assembly

From the control perspective, it is usually expected to find an optimal current waveform to maximize the output torque and efficiency of a motor under prescribed constraints. Jahns and Soong [8] introduced a control-based technique for minimizing torque ripples by tuning the current waveform with an on-line or off-line controller. Chan et al. [9] and Kim et al. [10] proposed their sensorless drives for high bandwidth torque control by both the back EMF estimation with fundamental excitation and the saliency-based method with carrier frequency excitation. As for the driving patterns, the optimal current waveform to produce a maximum torque is proportional to the magnetic flux variation in the air gap, which shares the same waveform with the back EMF [11].

Among the previous researches, few papers have ever integrated both the design and control perspectives into the motor design. This paper initiates a systematic approach for designing a torque motor to improve its torque capacity and to reduce torque ripples from both the design and control points of view. The flowchart of the proposed design process is presented in Table 1.

SPECIFICATIONS

The proposed torque motor will operate for machine tools over a low speed range with high accuracy and resolution, such as the application to a multi-axis highprecision machine center, computer numerical control milling machine or semiconductor handling equipment. A radial-flux SPM motor is chosen as the target sample to satisfy the major specifications in Table 2.

PRELIMINARY DESIGN

According to the design specifications, a preliminary model of the SPM motor is proposed in Figure 1, featuring a large diameter-to-axial-length ratio, which provides a thin ring with more space for bearings, sensors and other components in the hollow shaft of the rotor. The outer surface of rotor is embedded with permanent magnets to face their surrounding stator teeth, and the exterior of the stator back iron is enclosed by a water cooling house.

Basic Motor Configuration

Winding Type

Concentrated windings have simpler structure with shorter end windings than distributed windings, thereby yielding less flux leakage and power loss. It is also possible for the motor of concentrated windings to increase torque production by similar numbers of slots and poles, which also result in significantly low torque ripples [12]. Therefore, concentrated winding is preferred and selected for the torque motor in this paper.

Determination of the Number of Slots and Poles

It has been proved that a motor of fractional slot and pole ratio with concentrated windings may have a high torque density, and a motor of similar numbers of slots and poles will make the torque ripple so low that neither rotor nor stator skewing is necessary. Three useful factors will help a designer determine the number of slots and poles:

1) Number of slots per pole per phase

$$N_{spp} = \frac{N_s}{2pN_{ph}} \quad (1)$$

where N_s is the number of slots, p is the number of pole pairs and N_{ph} is the number of phases. A motor is called fractional when N_{spp} has a fractional part; when N_{spp} is less than one and $N_s = 2p\pm2$, the motor is said to have similar numbers of slots and poles. Both properties render a motor with fewer torque harmonics, hence fewer ripples.

Table 1: Systematic design and control procedure

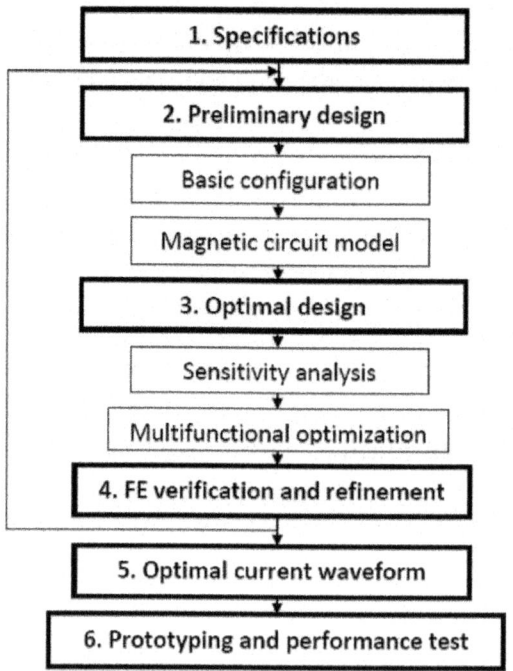

Table 2: Specifications of torque motor

Item	Specification
Objective	Torque motor for machine tools
Rated parameters	(1) Torque 518 Nm continuous (convection cooling) 1011 Nm continuous (water cooling) 1250 Nm peak (2) Voltage: 300-600 VDC (3) Current: 50-100 Amp (4) Speed: Rated 189 rpm (convection cooling), 333 rpm (water cooling)
Geometry	Outer diameter : 382 mm, Inner diameter: 240 mm, Motor length: 155 mm
Environment	Stator and rotor operation temperature 15-25 °C Coil winding operation temperature limited at 100 °C

Table 3: Fractional N_{spp} and winding factor k_w for 3-phase motors

$2p / N_s$	\multicolumn{10}{c}{N_{spp}/k_w}
	2	4	6	8	10	12	14	16	20	22
3	1/2 .866	1/4 .866	-	-	-	-	-	-	-	-
6	-	1/2 .866	-	-	1/5 .500	-	1/7 .500	1/8 .866	1/10 .866	-
9	-	-	1/2 .866	3/8 .945*	3/10 .945*	1/4 .764	3/14 .473	3/16 .328	3/20 .328	3/22 .617
12	-	-	-	1/2 .866	2/5 .933	-	2/7 .933	1/4 .866	1/5 .500	2/11 .250
15	-	-	-	-	1/2 .866	1/2 .866	5/14 .951*	5/16 .951*	1/4 .866	5/22 .711
18	\multicolumn{5}{c}{Distributed windings}	1/2 .866	3/7 .902	3/8 .945	3/10 .945	3/11 .945
21							1/2 .866	7/16 .890	7/20 .953*	7/22 .953*
24								1/2 .866	2/5 .933	4/11 .949

*Not recommended for unbalanced radial force

2) Winding factor k_w

The winding factor of concentrated windings is defined as

$$k_w = |\vec{E}_{phase}| N_{ph} / N_s \qquad (2)$$

where $\vec{E}_{phase}$ is the resultant back EMF phasor of a phase formed by its corresponding winding elements. The winding factor is an indication of what portion of the magnet is covered by the stator windings of a single phase.

3) Index factor C_T

The index factor C_T is defined to evaluate the amplitude of cogging torque for various slot and pole numbers:

$$C_T = 2pN_s / N_c \qquad (3)$$

where N_c is the least common multiple of the numbers of poles and slots. In general, a small index factor may indicate small amplitude of cogging torque; its smallest number is 1 when 2p and N_s are relatively prime.

It is convenient to make a table of N_{spp} and k_w for various combinations of slots and poles, among which the promising ones are selected in regard to the index factor, motor size and the performance of torque and torque ripple. Table 3 shows selected slot and pole combinations of fractional N_{spp}. First, those combinations with similar numbers of slots and poles, balanced windings and greater slot numbers are selected and marked in grey. Then, the largest winding factor 0.949 for the machine of 24 slots and 22 poles, designated the 24/22 machine, or other machines of the same slot and pole ratio, such as the 48/44 machine, becomes one of the candidates for optimal design. Since the specification of diameter in Table 2 allows a large number of slots and poles, the 48/44 machine is selected.

To validate this choice, the performance test was made by comparing it with other arrangements of slots and poles with winding factors of 0.866 and 0.933, which belong to the classes of the slot and pole ratios of 3:2 and 6:5, respectively. Since the diameter specification allows the inclusion of more slots, the machines of 72/48, 60/50 and 48/44 were selected for the performance test in terms of torque and torque ripple, as shown in Table 4.

In this test, all machines were excited with the same magneto-motive force (MMF) from magnets and the same phase current of 50 a from stator windings. The 48/44 machine produced the highest torque among the three machines. Although the torque ripple of the 48/44 machine was a little higher than the 60/50 machine, because of its flux saturating in the stator teeth, it is still the best choice among the three for its largest torque, winding factor and smallest index factor. Being configured of optimal design and finite element refinement will further improve its performance.

Determination of Magnet Shape

The back EMF waveform of SPM motors depends on the shape and pitch of magnet poles and stator teeth. The rectangular magnet is most commonly used because of certain manufacturing and cost advantages. The fan-shaped magnet is an improvement over the rectangular one, but it may cause manufacturing complexity.

Table 4: Performance test result

Parameter	Value		
Slot/Pole	72/48	60/50	48/44
Winding factor	0.866	0.933	0.949
Index factor	24	10	4
Average torque	1022 Nm	1046 Nm	1095 Nm
Torque ripple	12.8%	4.1%	6.6%

Table 5: Rated specifications for different magnet shapes

Specification	Magnet shape		
	Rectangular	Fan	Arc
Magnet weight (kg)	6.18	6.32	5.56
Rated torque at 64 A (Nm)	1177	1182	1118
Ripple range (Nm)	57.7	50.6	40.9
Active torque density (Nm/kg)	190	187	201
Magnet properties: Remanence (B_r) is 1.23T and relative permeability (μ_r) is 1.1			

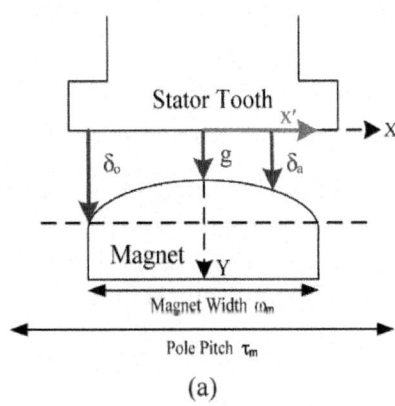

(a)

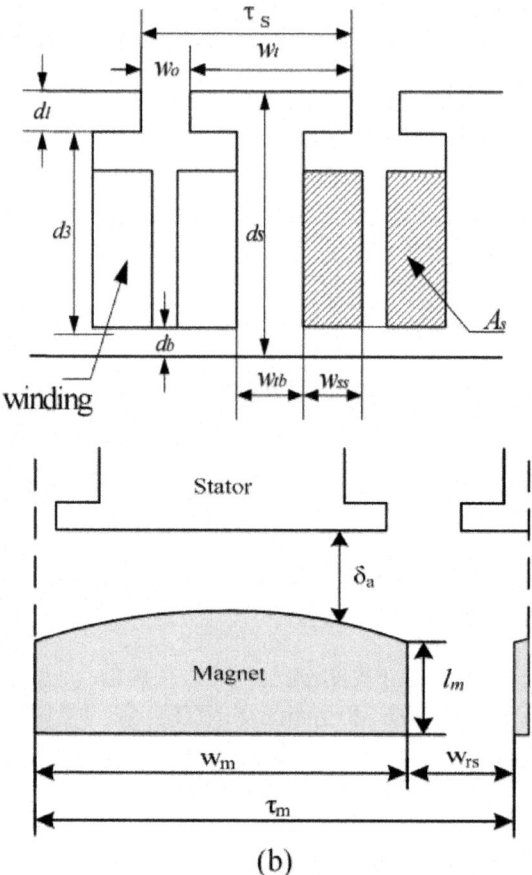

Figure 2: (a) Arc magnet and (b) design variables of stator and rotor of SPM machine

The arc magnet is promising for producing a smooth sinusoidal back EMF so as to match the excitation of the sinusoidal current waveform. An ideal arc shape in Figure 2(a) is expressed as a cosine function of magnet pole pitch t_m [5] and the air gap length d_r on the rotor side can be approximated by

$$\delta_r \approx \delta_a - g = [\frac{1}{\cos(\pi x'/\tau_m)} - 1]g, \text{ for } |x' - n\tau_m| < w_m/2,$$
$$\delta_r = \delta_0 - g, \text{ for } w_m/2 < |x' - n\tau_m| < \tau_m/2, n = 1,2,3... \quad (4)$$

where d_0 denotes the air gap length between the top of the stator teeth and the bottom of the rotor. The coordinates x and x' are the peripheral coordinate fixed on the stator and rotor, respectively, along the circle of the average radius $R = (R_{si} + R_{ro})/2$, while R_{si} is the inner radius of the stator and R_{ro} is the outer radius of the rotor.

Table 5 compares the motor performance for different magnet shapes. The motor with fan magnets provides the maximum rated torque at 64 A, but has a minimum torque density for extra magnet weight. However, the range of ripples, measuring the difference between the upper and lower values of torque ripple, is minimal for the motor with arc magnets. Its active torque density, defined as the ratio between the rated torque and magnet weight, is also maximal among the three. Besides, the back EMF constant produced by the motor with arc magnets is closest to a sinusoidal function, thus providing the lowest torque ripple. The arc magnets, therefore, become the best choice at the preliminary design stage.

Analytical Magnetic Circuit Model

According to the preliminary motor configuration, a 2D analytical magnetic circuit model based on the theory of electromechanical energy conversion is built to describe the performance of the 3D motor in terms of a set of objective functions, such as motor torque, torque density, speed and efficiency. These objectives are expressed as functions of motor geometries, which are illustrated as design variables in Figure 2(b) and are to be determined through the optimization process.

In addition to the assumptions of material linearity and the collinearity of flux and field densities, it is also necessary to make three additional assumptions for the 2D magnetic circuit model:

- The motor is operated in the linear range of the B-H curve of the magnetic material.
- The air-gap reluctance of the slotted stator structure is approximated by the effective air-gap length with Carter's coefficient [13].
- The flux flows straight across the air gaps between the stator and rotor, ignoring the fringing flux to simplify the analysis.

Therefore, the 3D motor structure in Figure 1 can be approximated by a 2D configuration inFigure 3(a) to facilitate the magnetic circuit analysis. By neglecting the flux leakage and armature reaction between the stator and rotor, the MMF from stator windings F_s is simply a square function of magnitude $N_t I$; while the MMF from rotor magnets F_r is a square function of magnet $H_c l_m$, where H_c and l_m are the coercivity and length of magnet, respectively. The overall MMF distribution is a linear combination of the MMFs of stator windings and rotor magnets

$$F(x,s) = F_s(x,s) + F_r(x,s) \qquad (5)$$

where s denotes the rotor shift, which is defined as the relative angle between the rotor and stator.

The magnetic flux density in the air gap is described as

$$B_g = \mu_0 \frac{F(x,s)}{\delta(x,s)} \qquad (6)$$

where m_0 is the permeability of air, d (x,s) is the effective air gap as a function of slot opening and slot pitch, as shown in Figure 3(b). By the use of the detailed expression of effective air gap [13], the field coenergy in the air gap is expressed as

$$W_c(s) = \int FB_g dA = L\mu_0 \int_0^{2\pi R} \frac{F^2(x,s)}{\delta(x,s)} dx \qquad (7)$$

where L is the axial length of the motor. The torque produced in the motor is then obtained by calculating the variation of magnetic coenergy in the air gap with respect to the rotor shift:

$$T(s) = R \frac{\partial W_c(s)}{\partial s}\bigg|_{I=constant} \qquad (8)$$

It may be possible to express the coenergy and torque in analytical forms. However, the numerical analysis is used instead to get torque distribution from such a complicated magnetic model. Therefore, the coenergy and torque (7) and (8) are rebuilt in the discrete form as

$$W_c(s) \approx W_c(h) = \mu_0 L \Delta\theta \sum_{k=1}^{N} \frac{F^2(k\Delta\theta, h\Delta\theta)}{\delta(k\Delta\theta, h\Delta\theta)} \qquad (9)$$

$$T(s) \approx T(h) = [W_c((h+1)\Delta\theta) - W_c(h\Delta\theta)]/\Delta\theta\big|_{I=constant}, h=1,2,3,\ldots \qquad (10)$$

where each electric period is divided into N equally spaced points, separated with mechanical position $\Delta\theta$ from each other, while $x = k\Delta\theta$ and $s = h\Delta\theta$. These discrete equations will be used for the multi-functional optimal design in the next step.

OPTIMAL DESIGN

The performance of the torque motor is usually evaluated by its maximum torque, torque density and efficiency, which are also known as objective functions or performance indices, describing the mechanical and electrical dynamics in terms of motor geometries, magnetic materials and driving conditions, as follows:

1) Rated torque

$$T_{r\,max} = \text{average of } R \frac{\partial W_c(s)}{\partial s}\bigg|_{I=rated\ current} \qquad (11)$$

2) Torque density

max $T_d = T_r / W$ (12)

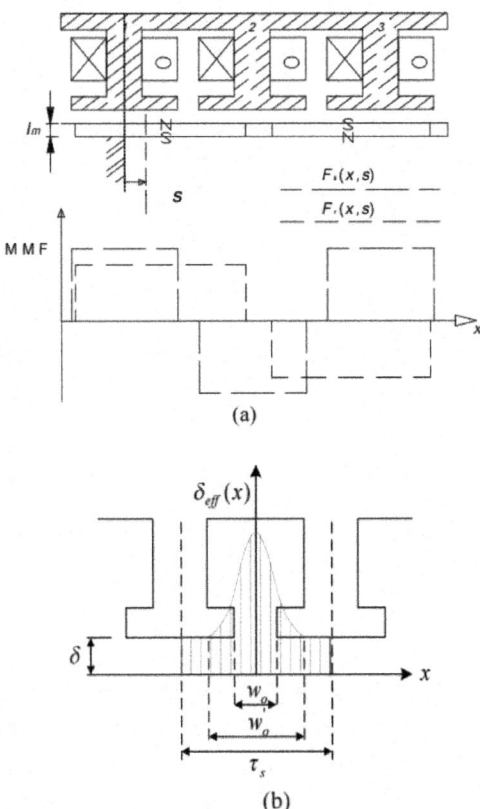

Figure 3: (a) 2-D motor configuration and magneto-motive force, and (b) effective air gap width due to slotting effect

3) Rated efficiency

max $\eta_r = \dfrac{T_r \omega_r}{T_r \omega_r + P_r + P_e + P_s}$ (13)

Here, the rated torque T_r is an implicit function of design variables, and is calculated through analytical magnetic circuit models. The motor weight W, rated speed w_r and rated efficiency h_r are all explicit functions of design variables. Their discrete equations can be easily derived and expressed similarly to (9) and (10).

Sensitivity Analysis

First, sensitivity analysis is required to determine the derivatives of the

objective functions with respect to the parameters of interest, then a set of design variables are determined. The purposes of sensitivity analysis are:
- The designer may want to discard those design variables with the least sensitivity of torque, torque density, torque ripple and/or machine efficiency.
- The designer may keep those design variables constant with sensitivities that are linear, or monotonic functions.
- Only those design variables that are not included in the above two cases are retained for the subsequent optimal design.

Table 6 lists all the variables for the sensitivity analysis, while other motor parameters are predetermined in Table 7 according to physical facts and previous design experiences. It would be a time and space-consuming process to illustrate all the sensitivity curves, though it is worth defining sensitivity indices as the ratio of the variation of motor performance and the variation of design variable. For example, the sensitivity index of maximum torque is denoted by $\Delta T_r / \Delta \varphi$, where φ symbolizes for design variables. Similarly, the sensitivity indices of toque density T_d and efficiency h_r are expressed, respectively, as $\Delta T_d / \Delta \varphi$ and $\Delta \eta_r / \Delta \varphi$. These sensitivity indices are easily normalized with their maximum values as denoted on the x-axis of Figure 4.

It is found that the output torque is greatly affected by the variations of air-gap length (δ), tooth width (w_{tb}), slot opening (w_o), magnet length (l_m) and magnet fraction ($a_m = w_m / \tau_m$) as shown in Figure 4(a). However, the back iron thickness (d_b), number of turns (N_t) and the copper wire diameter (d_w) have little influence on the output torque. Figure 4(b) indicates that the torque density is greatly influenced by the magnet fraction, magnet length and slot opening while other variables seem to be unrelated to torque density. Figure 4(c) shows that the magnet length, slot opening and air-gap length, and, especially, the magnet fraction, have significant influence on the torque ripple. Other variables, such as tooth width, shoe depth (d_1), back iron thickness (d_b), number of turns and copper diameter, seem to have no influence on the torque ripple. According to the sensitivity index of efficiency, it is not surprising that both slot opening and tooth thickness have an indispensable influence on efficiency as shown in Figure 4(d). Furthermore, both the air-gap length and the diameter of copper have significant influence on the motor efficiency.

According to the sensitivity indices, the shoe depth, back iron thickness and copper diameter can be discarded because they do not significantly influence the performance of the motor. The air-gap length, which is a compromising factor between motor performance and cost, is set at 1 mm by considering the manufacturing tolerance and accuracy. Finally, five design variables —

the magnet fraction, magnet length, slot opening, tooth thickness and number of turns, denoted by stars in Table 6 - are determined for multifunctional optimization.

Multifunctional Optimization

In terms of the design variables chosen from the sensitivity analysis, the performance indices — torque, torque density and efficiency — of the torque motor are implicitly or explicitly written as (11) through (13). The compromise programming method in the multifunctional optimization system tool (MOST) [14] is applied to search for the optimal values of design variables that maximize these performance indices.

The optimizer weights the performance indices to reach a satisfactory compromise among the design variables subject to the prescribed constraints: 1) The motor dimensions must be realized, e.g. $R_o > R_i$; 2) The slot opening is twice larger than the air gap length, but 0.35 times less than the slot pitch; 3) The shoe depth fraction, defined as the ratio between shoe depth (d_1) and tooth length (d_s), is confined less than 0.5; 4) The slot current density is less than 9×10^6 A/m^2; 5) The flux density in the electrical steel is less than its saturation value of 1.8 T; 6) The peak value of back EMF per phase should be less than the component of the driving voltage along the back EMF vector.

Different weightings were assigned to the three performance indices for which relative importance is addressed for the optimization, but the three best results in terms of motor performances are listed and compared in Table 8. The example with the highest weighting on efficiency, column 1:1:6, cannot provide enough torque and torque density, and is eliminated first.

Table 6: Design variables for sensitivity analysis

No.	Name	Variable
1	Magnet fraction*	a_m
2	Magnet length*	l_m
3	Slot opening*	w_o
4	Shoe depth	d_s
5	Tooth width*	w_{tb}
6	Stator back iron thickness	d_b
7	Air gap length	δ
8	Number of turns*	N_t
9	Copper wire diameter	d_w

*Candidates for optimal design

Table 7: Predetermined variables for sensitivity analysis

Specifications	Value
Stator outer radius	191 mm
Rotor inner radius	120 mm
Rotor axial length	155 mm
Magnet shape	Arc
Relative permeability	1.1 H/m
Magnet density	7.45 g/cm^3
Magnet remanence	1.23 T
No. of Phase	3
Current waveform	Sinusoidal wave
No. of slots	48
No. of poles	44
No. of turns	60
Magnet length	8 mm
Slot opening	4 mm
Tooth width	11.5 mm
Air gap length	1 mm
Magnet fraction	0.8
Length of stator tooth	38.04 mm
Stator back iron thickness	10 mm
Shoe depth	5 mm
Copper wire diameter	1 mm

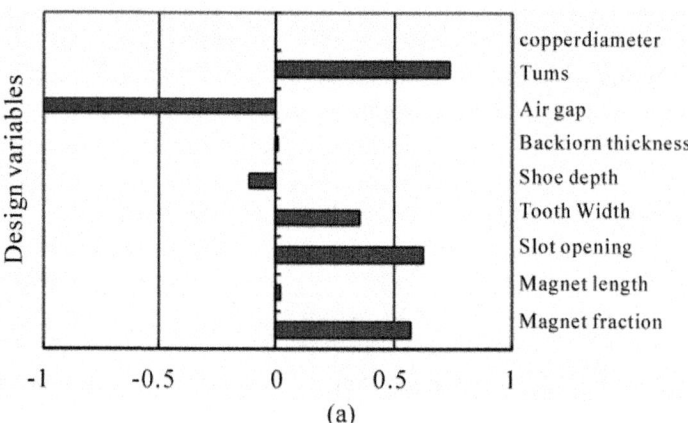

(a)

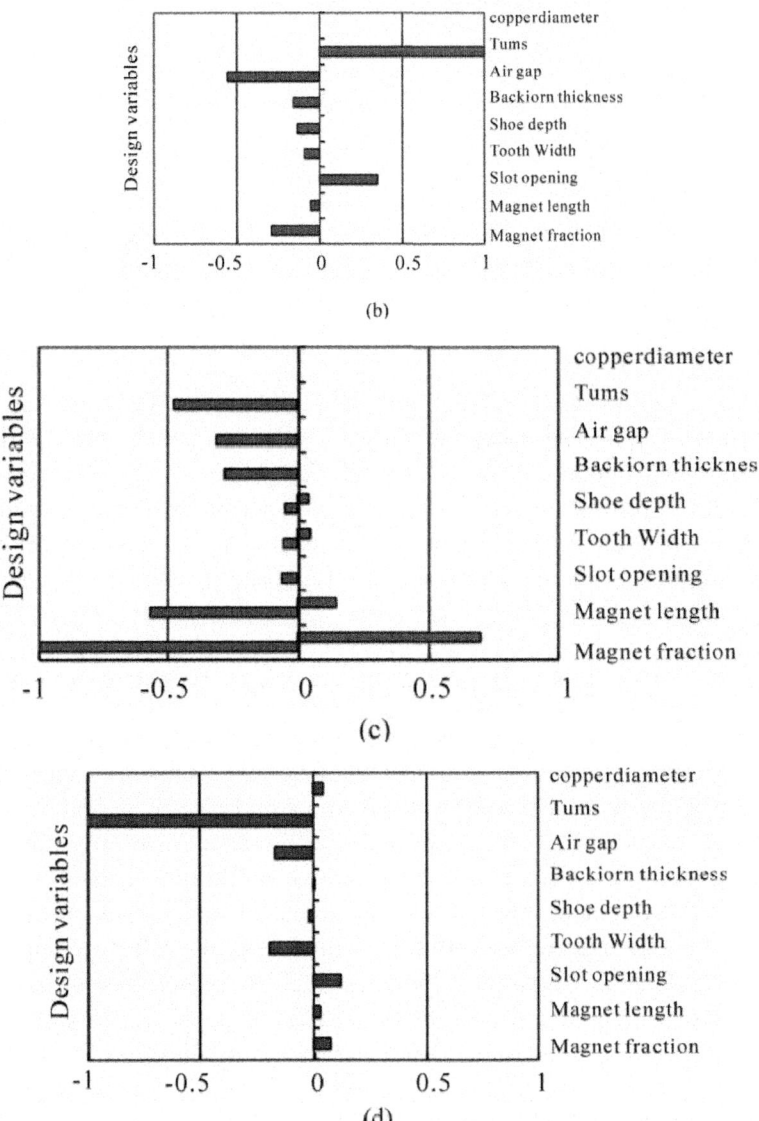

Figure 4: Sensitivity index of (a) output torque, (b) torque density, (c) torque ripple, and (d) efficiency

Table 8: Optimized motor variables and performance

$T_r : T_r/W : \eta_r$	1:1:1	6:1:1	1:1:6
Stator slot opening(mm)	4.098	4.979	5
Tooth thickness(mm)	11.41	12.01	11.11
Magnet length (mm)	7.98	8	6.56
Magnet fraction	0.823	0.871	0.808
Number of turns	61	65	59
Phase current (A_{rms})	50	50	50
Motor weight (kg)	58.42	58.62	57.35
Ohmic loss (W)	3089	3937	2073
Core loss (W)	98	98.7	92
Maximum speed (rpm)	132	116	143
Output torque (Nm)	1091	1240	975
Torque density (Nm/kg)	18.34	20.70	16.68
Efficiency (%)	90.02	87.70	92.8

The example of column 6:1:1 with the highest weighting on torque has the highest torque and torque density, but yields the worst performance in ohmic and core loss, maximum speed and efficiency among the three examples. Finally, the optimization result with equally important weightings on the three objectives, column 1:1:1, presents a balanced and satisfactory performance in torque, torque density and efficiency, and is chosen for the FE verification and refinement.

FINITE ELEMENT VERIFICATION AND REFINEMENT

The 3D FE analysis is responsible for verifying the effectiveness of the 2D optimal design by including nonlinear characteristics. Owing to the symmetric structure, a quarter section of the motor with a half electric period is sufficient to model and analyze the expected performance. The resulting motor performance by the FE verification is shown in Table 9. It is also expected that the FE refinement through the adjustment of motor geometries, such as shoe depth, slot opening and back iron thickness, must further improve the motor performance. Apparently, the torque density is increased because of the reduction of weight, and the torque ripple is also suppressed by reshaping the geometry of slot opening, tooth depth and back iron thickness.

OPTIMAL CURRENT WAVEFORM

It has been proved that the maximum torque is produced if the phase current is proportional to the angular rate of change of the field flux of the motor [15]. In other words, the best current waveform is the same as the back EMF wave of the motor in order to produce the maximum torque with minimal current. The reduction of torque ripple in the previous design steps means the back EMF of the motor needs to be made as close as possible to a sinusoidal wave. Besides, the current phase shift may have some influence on the torque production especially for the configuration of similar numbers of slots and poles in the three-phase drive. Figure 5(a) illustrates the minimum magnetic circuit model with 12 slots and 11 poles for the 48/44 machine, where the coils are arranged as AA'AA'B'BB'BCC'CC', and coils A and A' are wound in opposite directions. Figure 5(b) lays out the EMF vectors of windings, where the slots of coils A_1, $-B_1$, and C_1 are distributed in 120°E (electric degrees) of phase offset and the best shift g of each phase current must lead them in 22.5°E, thereby remaining a balanced configuration. It was found that not only the torque increased 8.4%, but the ripple also decreased by changing the current shift from 0 to 22.5°E.

PROTOTYPE FABRICATION AND EXPERIMENTS

The axial length of the motor was previously specified at 155 mm. To make the prototyping easier, it was downsized to 55 mm in view of precise machining and accurate assembling. Inevitably, such reduction of axial length must cause the rated torque to decrease by about 1/3 of the optimally designed motor. Figure 6(a) shows the stator core and concentrated windings, and the rotor assembly, glued on the outer surface of which are NeFeB 40SH magnets. A complete assembly of the torque motor is shown in Figure 6(b). The performance of the prototype motor was tested with the voltage supply of 220V. Figure 7 compares the back EMF waveforms from the FE analysis and experiment, where both curves are close to the pure sine function, but the error between the experimental back EMF and pure sine function is even less than 5%. It is also interesting to point out that the back EMF constant from the experiment is 4.48 V/rad/s, which is very close to 4.57 V/rad/s from the FE analysis. Figure 8 shows the relationship between the torque and current, where $T_{mag}=31.2I_{rms}$ is for the magnetic circuit model, $T_{FE}=30.7I_{rms}$ for the FE analysis and $T_{exp}=33.7I_{rms}+5.43$ for the experiment. The measured torque constant 33.7 Nm/A from the experiment is close to but slightly higher than those from the magnetic circuit model and FE analysis, partially because the measured torque may include additional friction from the bearing of the motor assembly. The offset of 5.434 Nm may account for the friction in bearing or inevitable measurement errors.

SUMMARY AND CONCLUSIONS

A systematic approach of design and control for a permanent magnet synchronous torque motor for machine tools was proposed. The design procedure was illustrated step by step in the order of specification, preliminary design, optimal design, finite element verification and refinement, and optimal control current waveform. These procedures help an engineer perform a complete design without missing any key issues. The sensitivity analysis and multifunctional optimization provide efficient and essential information for the designer to make a final decision for motor geometry. The precise finite element tool not only verifies but also refines the preliminary and optimal design of the machine by improving its performance. The determination of optimal current waveform and phase shift further allows the machine to operate at its best capacity. Finally, the experiments on a prototype presented satisfactory performance in terms of back EMF wave, back EMF constant and torque constant, thereby validating the effectiveness of the optimal design and control strategy. The proposed design approach must facilitate motor designs in an effective way, especially for an engineer of little experience. The only deficiency of this study is the necessity of creating magnetic circuit models which differ from motor types and are only best approximated in two dimensions under various linear assumptions.

Table 9: FE verification and refinement

Variables	FE verification	FE refinement
Slot opening (mm)	4.098	4.210
Stator back iron (mm)	10	7
Tooth width (mm)	11.41	11.50
Rated torque (Nm)	1081	1122
Motor weight (kg)	60.67	57.96
Torque density (Nm/kg)	18.15	18.93
Torque ripple (%)	3.6	2.29
Copper ohmic loss (W)	3084	3084
Core loss (W) (at rated speed)	96.2	90.0
Efficiency (%)	90.0	90.3

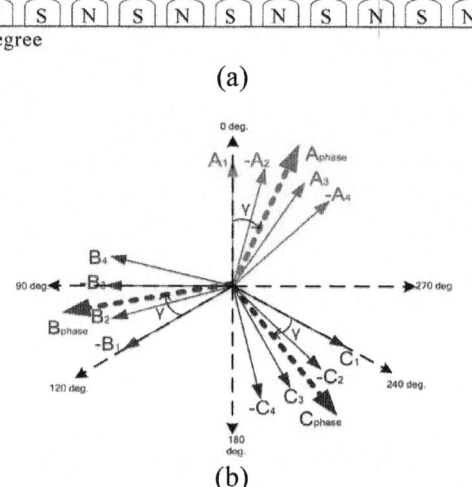

Figure 5. (a) Minimum magnetic circuit model of the 48/44 machine, and (b) its balanced winding layout

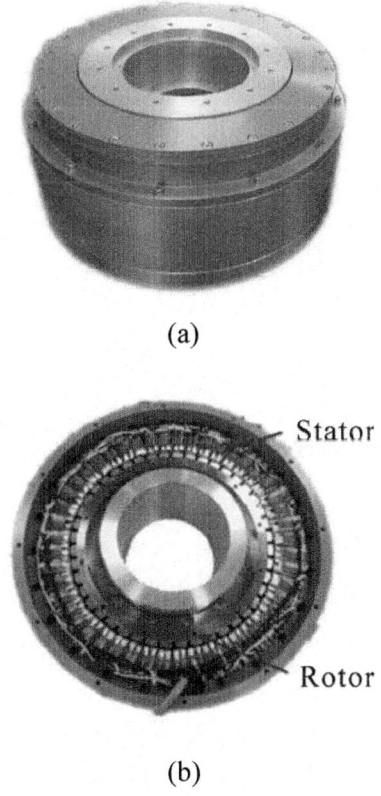

(a)

(b)

Figure 6. (a) Stator, rotor, and (b) their assembly of motor prototype

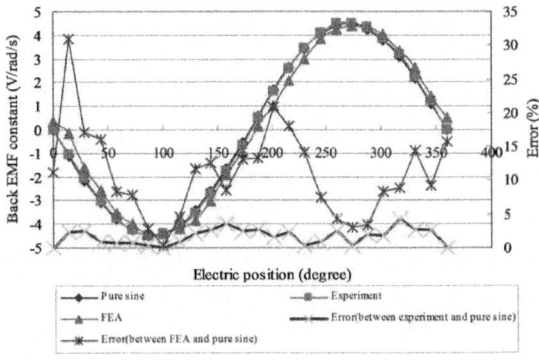

Figure 7: Back EMF waves from FE analysis and experiment

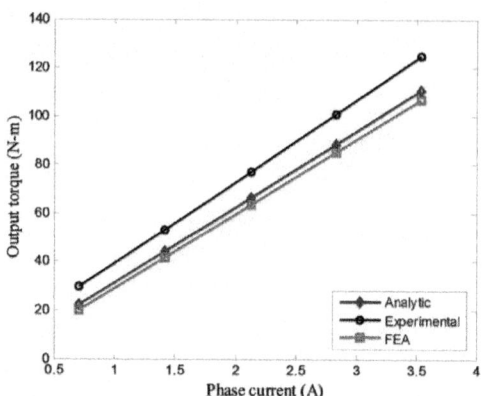

Figure 8: Torque constant curves

ACKNOWLEDGMENTS

This work was supported by National Science Council of Taiwan, China, under Contract NSC95-2221-E-002-132- MY2.

REFERENCES

1. B. Stumberger, G. Stumberger, M. Hadziselimovic, A. Hamler, M. Trlep, V. Gorican, and M. Jesenik, "Highperformance permanent magnet brushless motors with balanced concentrated windings and similar slot and pole numbers," Journal of Magnetism and Magnetic Materials, Vol. 304, pp. e829–e831, 2006.
2. D. Ishak, Z. Q. Zhu, and D. Howe, "Comparison of PM brushless motors, having either all teeth or alternate teeth wound," IEEE Transactions on Energy Conversion, Vol. 21, pp. 95–103, 2006.
3. C. C. Hwang, S. P. Cheng, and C. M. Chang, "Design of high-performance spindle motors with concentrated windings," IEEE Transactions on Magnetics, Vol. 41, pp. 971–973, 2005.
4. J. Cros and P. Viarouge, "Synthesis of high performance PM motors with concentrated windings," IEEE Transaction on Energy Conversion, Vol. 17, pp. 248–253, 2002.
5. M. F. Hsieh and Y. S. Hsu, "An investigation on influence of magnet arc shaping upon back electromotive force waveforms for design of permanent-magnet brushless motors," IEEE Transactions on Magnetics, Vol. 41, pp. 3949–3951, 2005.
6. M. S. Islam, S. Mir, T. Sebastian, and S. Underwood, "Design

considerations of sinusoidally excited permanent-magnet machines for low-torque-ripple applications," IEEE Transactions on Industry Applications, Vol. 41, pp. 955–962, 2005.
7. Y. P. Yang, W. C. Huang, and C. W. Lai, "Optimal design of rim motor for electric powered wheelchair," IET Electric Power Applications, Vol. 1, pp 825–832, 2007.
8. T. M. Jahns and W. L. Soong, "Pulsating torque minimization techniques for permanent magnet AC motor drives-a review," IEEE Transactions on Industrial Electronics, Vol. 43, pp. 321–330, 1996.
9. T. F. Chan, W. Wang, P. Borsje, Y. K. Wong, and S. L. Ho, "Sensorless permanent-magnet synchronous motor drive using a reduced-order rotor flux observer," IET Electric Power Applications, Vol. 2, pp. 88–98, 2008.
10. T. Kim, H. W. Lee, and M. Ehsani, "Position sensorless brushless DC motor/generator drives: Review and future trends," IET Electric Power Applications, Vol. 1, pp. 557–564, 2007.
11. Y. P. Yang and D. S. Chung, "Optimal design and control of a wheel motor for electric passenger cars," IEEE Transactions on Magnetics, Vol. 43, 2007, pp. 51–61.
12. Z. Q. Zhu and D. Howe, "Influence of design parameters on cogging torque in permanent magnet machines," IEEE Transactions on Energy Conversion, Vol. 15, 2000, pp. 407–412.
13. V. Ostovic, "Computer-aided analysis of electric machines," New York: Prentice Hall, 1994.
14. C. T. Tseng, W. C. Liao, and T. C. Tang, "MOST user's manual," in Mechanical Engineering, 1.2ed Taiwan, Hsinchu: National Chiao-Tung University, 1993.
15. Y. P. Yang, Y. P. Luh and C. H. Cheung, "Design and control of axial-flux brushless dc wheel motors for electric vehicles – Part I: multi-objective optimal design and analysis," IEEE Transactions on Magnetics, Vol. 40, No. 4, July 2004, pp.1873–1882.

Chapter 4

ELECTRIC MOTOR PERFORMANCE IMPROVEMENT USING AUXILIARY WINDINGS AND CAPACITANCE INJECTION

Nicolae D.V

Tshwane University of Technology South Africa

INTRODUCTION

Generally, some electric machines such as induction machines and synchronous reluctance motors require reactive power for operation. While the reactive power required by a synchronous machine can be taken from the power source or supplied by the machine itself by adjustment of the field current, the power factor of an induction machine is always lagging and set by external quantities (i.e., the load and terminal voltage). Poor power factor adversely affects the distribution system and a cost penalty is frequently levied for excessive VAr consumption.

Power factor is typically improved by installation of capacitor banks parallel to the motor. If the capacitor bank is fixed (i.e. that it can compensate power factor only for a fixed load), when the load is variable, then the compensation is lost. Some authors (El-Sharkawi et al, 1984, Fuchs and Hanna, 2002) introduced the capacitors using thyristor/triac controllers; by adjusting the firing angle, the capacitance introduced in parallel with the motor becomes variable and thus compensating the power factor for any load. Other works (Suciu et al, 2000.) consider the induction motor as an RL load and power factor is improved by inserting a variable capacitor (through a bridge converter) which is adjusted for unity according with the load. For the above methods, the capacitive injection is directly into the supply. Another method conceived for slip ring induction motor was to inject capacitive reactive power direct into the rotor circuit (Reinert and Parsley, 1995; Suciu, et al. 2002). The injection of reactive power can be done through auxiliary windings magnetically coupled with the main windings (E. Muljadi *et al.* 1989; Tamrakan and Malik, 1999; Medarametla et al. 1992; Umans, and H. L. Hess, 1983; Jimoh and Nicolae, 2006, 2007). This compensating method has also been applied with good results not only for induction motors but also for a synchronous reluctance motor (Ogunjuyigbe *et al.* 2010).

METHOD DESCRIPTION

Physical Solution

The method described in this chapter makes use of two three-phase stator windings. One set, the main winding (star or delta), is connected directly to the source. The other set of windings - auxiliary, is only magnetically coupled to the main winding. All windings have the same shape and pitch, but may have different turn numbers and wire sizes; usually smaller in order to be accommodated in the slots together with the stator. The windings are arranged in slots such that there is no phase shift between the two windings. Figure 1 shows a possible arrangement of the windings for a four pole induction machine.

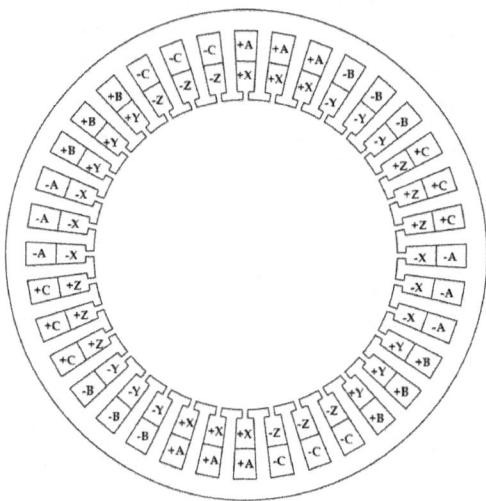

Figure 1: High Power factor induction machines-windings arrangement

Auxiliary Windings Connections

As mention above, the main winding can have delta or star connection. Figure 2 shows the main winding connected in star and the auxiliary windings connected in generic (a), star (b) and delta (c) to the capacitor bank via a static switch.

Figure 3 shows a simpler way to inject capacitive reactive power. In this method, the auxiliary windings are in "single –phase connection" with the apparent advantage of using only one capacitor and static switch.

Variable Capacitors

In order to achieve a compensation for various loading of the machine, the compensating capacitor should be able to be varied. This capability is obtained through connecting a fixed capacitor via a static switch. The static switch can be achieved using thyristors or IGBTs in bidirectional configuration.

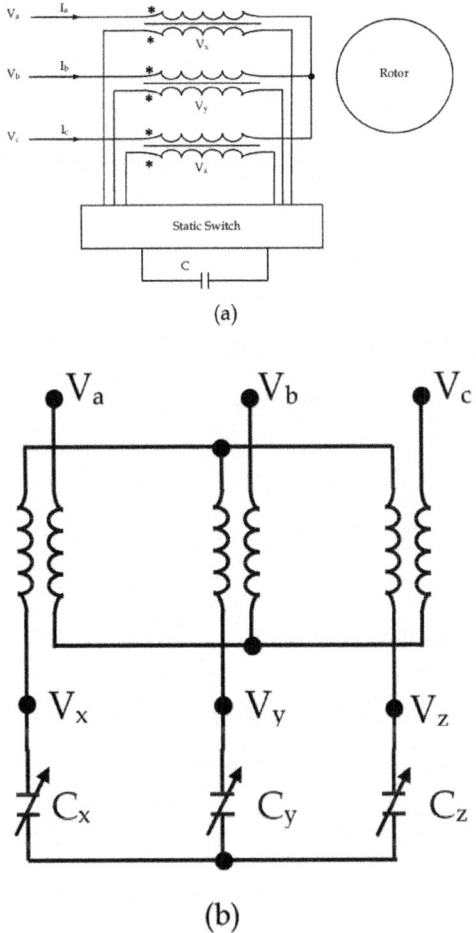

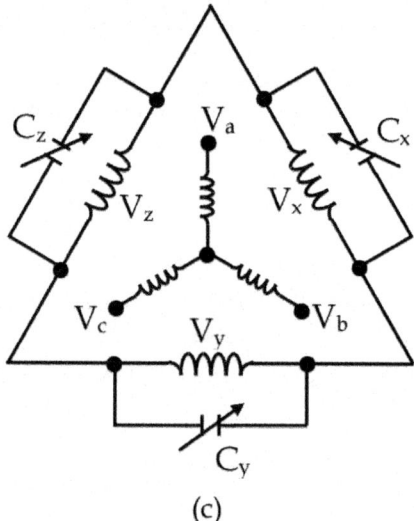

Figure 2: Auxiliary windings: a) generic connection; b) star connection; c) delta connection

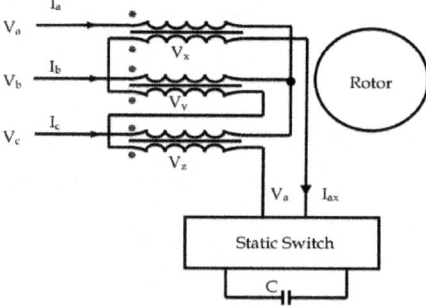

Figure 3: Auxiliary windings: "single –phase connection"

Thyristor-based variable capacitor

Figure 4 shows the use of thyristor to accomplish a variable capacitor. The inductor L_r is introduced to reduce – limit the surge current; it is relatively small and does not affect the overall capacitive behaviour.

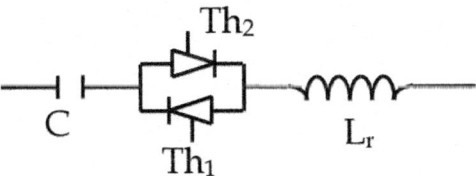

Figure 4: Variable capacitor using bidirectional thyristor

The equivalent capacitance depends on the delay angle. Due to the phase angle control, the device introduces harmonic currents.

IGBT-based variable capacitor

The above drawback can be address using IGBTs in bidirectional configuration (Figure 5) and a switching frequency higher then operational frequency (50 Hz).

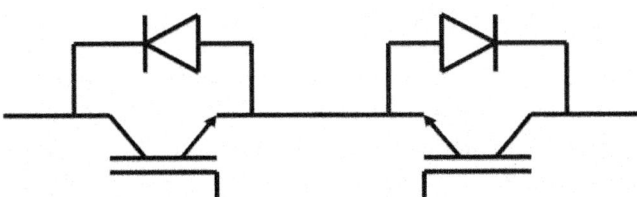

Figure 5: IGBT in bidirectional topology

Figure 6 shows a configuration to achieve a variable capacitor using two bidirectional static switches. The main capacitor C_1 is introduced in the auxiliary winding circuit, via a bidirectional switch Sw_1, for a period of time depending on the duty cycle ( ) of the switching frequency; in this time the bidirectional switch Sw_2 is OFF. When Sw_1 is OFF, the capacitor is discharged. The reactor L_r limits the capacitive surge current without affecting the capacitive behaviour.

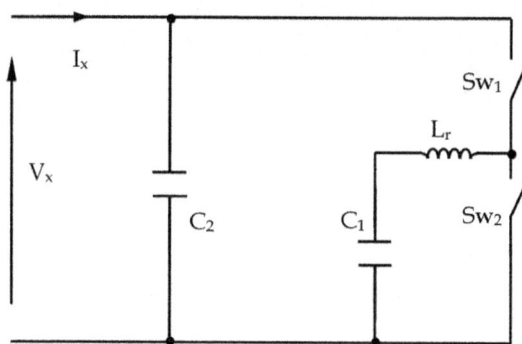

Figure 6: Variable capacitor using two IGBTs in bidirectional topology

The capacitor C_2, much smaller than C_1 is connected to mitigate the voltage spikes during switching off the main capacitor. Thus, the equivalent capacitor can be written as:

$$C_{eq} = \delta \times C_1 + C_2 \tag{1}$$

Variable capacitor H-topology

Figure 7 shows a single-phase H topology to achieve a variable capacitor. This configuration using H-bridge bidirectional topology obtains a higher equivalent capacitance for the same fixed one as reference. In this configuration, the reactor L_r has the same purpose of limiting the surge capacitive current, while C_2 also of small value mitigates the voltage spikes. The equivalent capacitance could be express as:

$$C_{eq} = C_2 + \frac{C_1}{(2\delta - 1)^2} \tag{2}$$

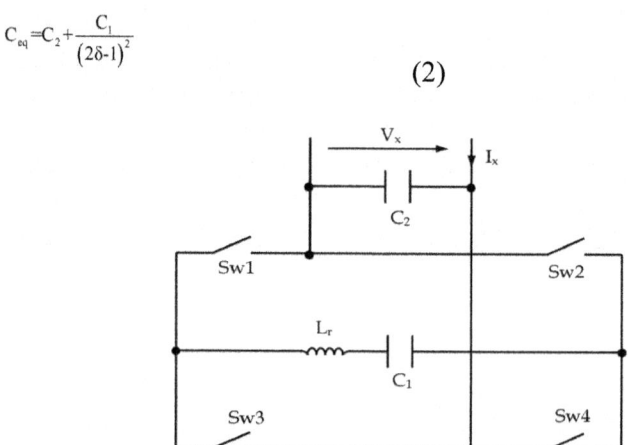

Figure 7: Variable capacitor using H-bridge bidirectional topology

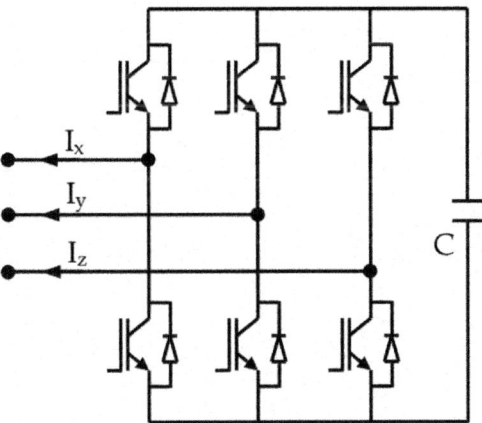

Figure 8: Variable capacitor using three-phase H-bridge topology

It can be notice that the equivalent capacitance could increase significant when the duty cycle approaches 50 %. In practice, the switches are not ideal and there is no "infinite increase" of the equivalent capacitance.

Another solution to achieve a variable capacitance, or rather to generate a capacitive current was proposed using a three-phase H topology as PWM inverter (E. Muljadi, *et al* 1989; Tamrakan and Malik, 1999) as presented in Figure 8. The converter injects capacitive reactive power into auxiliary windings and thus improving the power factor of the motor.

MATHEMATICAL MODEL

The machine is treated as having two three-phase windings and the voltages equations system can be written as:

$$[V_{abcs}] = [R_1][I_{abc}] + \frac{d}{dt}[\lambda_{abc}] \qquad (3)$$

$$0 = [R_2][I_{xyz}] + \frac{d}{dt}[\lambda_{xyz}] + Vc_{xyz} \qquad (4)$$

$$0 = [R_r][I_{abcr}] + \frac{d}{dt}[\lambda_{abcr}] \qquad (5)$$

$$V_{abc} = \begin{bmatrix} V_a & V_b & V_c \end{bmatrix}^T \qquad (6)$$

$$I_{abcs}=\begin{bmatrix} I_a & I_b & I_c \end{bmatrix}^T ; \quad \lambda_{abc}=\begin{bmatrix} \lambda_a & \lambda_b & \lambda_c \end{bmatrix}^T \qquad (7)$$

$$[R_1]=\begin{bmatrix} r_a & 0 & 0 \\ 0 & r_b & 0 \\ 0 & 0 & r_c \end{bmatrix} \quad [R_2]=\begin{bmatrix} r_x & 0 & 0 \\ 0 & r_y & 0 \\ 0 & 0 & r_z \end{bmatrix} \qquad (8)$$

Note that indices "1" refer to the main winding and "2" to the auxiliary winding.

$$\begin{bmatrix} \lambda_{abcs} \\ \lambda_{xyz} \\ \lambda_{abcr} \end{bmatrix} = \begin{bmatrix} L_{abc} & L_{abcsxyz} & L_{abcsr} \\ L_{xyzabcs} & L_{xyz} & L_{xyzabcr} \\ L_{abcrs} & L_{abcrxyz} & L_{abcr} \end{bmatrix} \begin{bmatrix} I_{abcs} \\ I_{xyz} \\ I_{abcr} \end{bmatrix} \qquad (9)$$

The inductances in eq. (9) are time dependent, and this make the equation difficult and time consuming to solve. In order to obtain constant parameters, the voltage equations (3-5) are then transformed to the rotor reference frame. To achieve this, the equations are multiplied with an appropriate transformation matrix $K(\theta)$ to obtain:

$$[K(\theta)][V_{abcs}]=[R_1][K(\theta)][I_{abcs}]+\frac{d}{dt}[K(\theta)][\lambda_{abcs}] \qquad (10)$$

$$0=[R_2][K(\theta)][I_{xyzs}]+\frac{d}{dt}[K(\theta)][\lambda_{xyzs}]+[K\theta][V_{cxyzs}] \qquad (11)$$

$$0=[R_r][K(\theta_r)][I_{abcr}]+\frac{d}{dt}[K(\theta_r)][\lambda_{abcr}] \qquad (12)$$

When these equations are expanded, after substantial matrix manipulations, it resolves to:

$$[V_{qdo1}]=[R_{s1}][I_{qdo1}]+\frac{d}{dt}[\lambda_{qdo1}]+\varpi[\lambda_{qdo1}] \qquad (13)$$

$$0=[R_2][I_{qdo2}]+\frac{d}{dt}[\lambda_{qdo2}]+\varpi[\lambda_{qdo2}]+[Vc_{qdo2}] \qquad (14)$$

$$0=[R_r][I_{qdor}]+\frac{d}{dt}[\lambda_{qd0r}]+\varpi_r[\lambda_{qdo1}] \qquad (15)$$

where

$$\omega = \omega \begin{bmatrix} 0 & 1 & 0 \\ -1 & 0 & 0 \\ 0 & 0 & 0 \end{bmatrix}; \omega_r = (\omega - \omega_r) \begin{bmatrix} 0 & 1 & 0 \\ -1 & 0 & 0 \\ 0 & 0 & 0 \end{bmatrix}$$

(16)

and

$$\left[Vc_{qdo2} \right] = \frac{1}{C} \int \left[I_{qd02} \right] dt + \omega \left[Vc_{qd02} \right]$$

(17)

Neglecting the '0' sequence since we initially are assuming a balanced system, the expression for the stator and rotor flux linkages, resolves into the matrix:

$$\begin{bmatrix} \lambda_{q1} \\ \lambda_{d1} \\ \lambda_{q2} \\ \lambda_{d2} \\ \lambda_{qr} \\ \lambda_{dr} \end{bmatrix} = \begin{bmatrix} L_{ls1}+L_m & 0 & L_{lm}+L_m & 0 & L_m & 0 \\ 0 & L_{ls1}+L_m & 0 & L_{lm}+L_m & 0 & L_m \\ L_{lm}+L_m & 0 & L_{ls2}+L_m & 0 & L_m & 0 \\ 0 & L_{lm}+L_m & 0 & L_{ls2}+L_m & 0 & L_m \\ L_m & 0 & L_m & 0 & L_{lsr}+L_m & 0 \\ 0 & L_m & 0 & L_m & 0 & L_{lsr}+L_m \end{bmatrix} \begin{bmatrix} I_{q1} \\ I_{d1} \\ I_{q2} \\ I_{d2} \\ I_{qr} \\ I_{dr} \end{bmatrix}$$

(18)

EQUIVALENT MODEL

Symmetrical loaded auxiliary windings

When each phase of the auxiliary windings is loaded with equal capacitors (C), eventually star connected, the entire circuit is having a symmetrical behaviour and the equivalent circuit is very simple as shown in Figure 9.

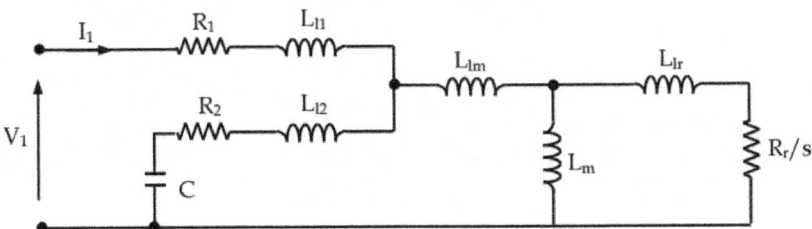

Figure 9: Equivalent circuit for symmetrical loading auxiliary winding

This equivalent circuit has two branches, each having separate resistance (R_1 – main and R_2 - auxiliary) and leakage reactance (L_{l1} – main and L_{l2} – auxiliary) together with a common mutual leakage inductance L_{lm}, which occurs due to the fact that the two set of windings occupy the same slots and therefore mutually coupled by their leakage flux. The mutual inductance that occurs between main winding and rotor is represented by L_m. Other parameters of the equivalent circuit are: main winding resistance R_1, auxiliary winding resistance R_2, main winding leakage reactance L_1, auxiliary winding leakage inductance L_2; mutual leakage inductance L_{1m}, rotor leakage inductance L_{1r}; and the rotor resistance R_r and "s" is the slip. In this analysis there is no need to refer the auxiliary winding quantities to the main winding, because the two sets of windings are wound for the same number of turns with a transformation ratio of one. The above equivalent circuit helps us to determine the input impedance seen from the supply and the condition for unity power factor.

$$Z_{in} = R_1 + jX_{11} + \frac{(R_2R_3 - X_2X_3) + j(X_2R_3 + X_3R_2)}{(R_2 + R_3) + j(X_2 + X_3)} \quad (19)$$

Where: R_1 and jX_{11} are the components of the main winding (per-phase impedance), X_2 is the equivalent reactance of the auxiliary winding including the compensating capacitor, R_2 is the resistance (per-phase) of the auxiliary winding, R_3 and X_3 are the equivalent resistance and reactance of the paralleling the rotor branch with magnetizing branch:

$$X_2 = \frac{1}{\omega C} - X_{12} \quad (20)$$

$$X_3 = \operatorname{Im}\{jX_{1m} + [(R_r/s) + jX_{1r}] // (jX_m)\} \quad (21)$$

$$R_3 = \operatorname{Re}\{[(R_r/s) + jX_{1r}] // (jXm)\} \quad (22)$$

The condition for unity power factor is:

$$\operatorname{Im}\{Z_{in}\} = 0 \quad (23)$$

Which, after some mathematical manipulation can be written as:

$$\alpha X_2^2 + \beta X_2 + \gamma = 0 \quad (24)$$

With:

$$\alpha = X_3 + X_{11} \quad (25)$$

$$\beta = -R_2 R_3 + (R_1 + R_2)R_3 + 2X_3 X_{11} \quad (26)$$

$$\gamma = -X_3\left[R_2R_3 - (R_2+R_3)R_2\right] + X_{11}\left[(R_2+R_3)^2 + X_3^2\right] \quad (27)$$

The equation (24) together with relations (25) to (27) produces two solutions, which means for a given slip (s) there are two values for the capacitor satisfying the condition for unity power factor. The practical/recommended value is the high X_2 or small C connected to the auxiliary winding in order to have a small current through it.

Asymmetrical loaded auxiliary windings

If the auxiliary windings are connected as "single-phase configuration", then the system has an asymmetrical behaviour. This connection is obtained by connecting in series the three auxiliary windings and thus we can write: $I_x = I_y = I_z = I_x$. Using this condition of current in the expansion that results to equations (3) – (18) it can be written: $I_{d2} = I_{q2} = 0$, thus the resulting expression from this can be represented in a d-q-0 equivalent circuit of Figure 10 shown below.

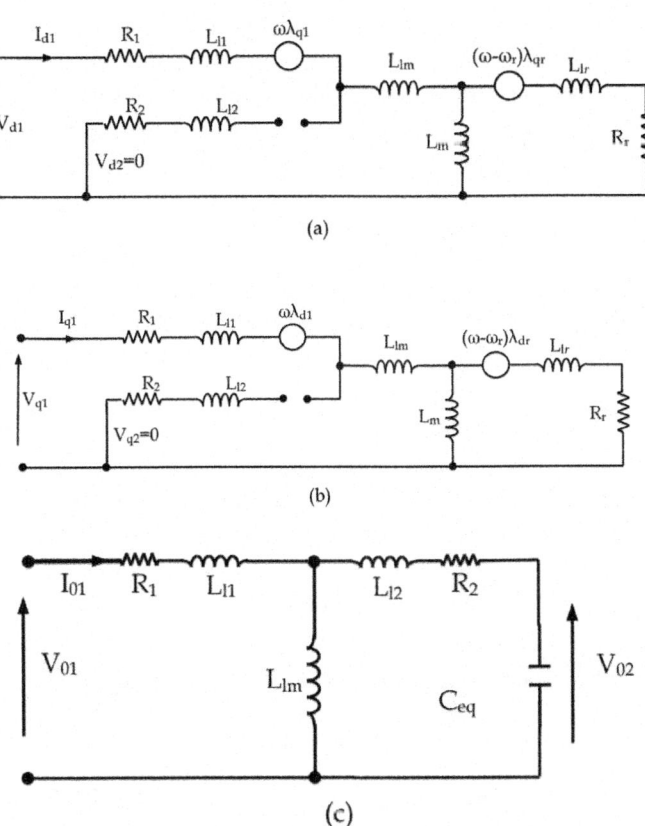

Figure 10: The "d-q-0" equivalent circuit: a) "d" – equivalent circuit; b) "q" – equivalent circuit; c) "0" – equivalent circuit

The particularity of connecting the auxiliary windings in a single phase winding creates an asymmetrical situation which brings about the relevance of the zero sequence. The power factor of the machine could be defined by the argument of Z_a, which is V_a/I_a. And this is expressed as:

$$Z_a = \frac{V_a}{I_a} = \frac{V_{q1}+V_{01}}{I_{q1}+I_{01}} \qquad (28)$$

$$V_{01}(C) = (-j\omega C + Z_{02}) \times I_{02} \times \frac{(Z_{01}+Z_{lm})}{Z_{lm}} \qquad (29)$$

As can be observed from the equations (28) and (29), the power factor of the machine depends on C and thus it can be brought to unity by means of adjusting the equivalent capacitance.

It should be also noticed that the asymmetrical behaviour of the auxiliary windings connected as "single-phase configuration" has got a significant drawback namely creating torque ripple. This disadvantage should not be overseen by the simplicity of the physical solution. Given this, further in this, we will consider only the symmetrical loading of the auxiliary winding.

CONCEPT VALIDATION

For the validation of this concept of improving performances of the induction motor injecting capacitive reactive power via auxiliary windings, a standard 4 kW, 380V, 50Hz, 4 pole, 36 slots with frame size DZ113A induction motor have been used. The stator was rewound with two full-pitches, single layer windings (see Figure 1).The main windings – delta connected have same number of turns like in the original motor but the size of the wire has been reduced to accommodate the auxiliary windings. The auxiliary windings were connected in delta in order to reduce current through them. For this study, the auxiliary windings have the same number of turns as main windings. The rotor is unchanged. Table 1 shows the parameters of the modified motor under test obtained using the IEEE test procedure. The mutual leakage inductance X_{lm} is very small compared to the other reactance's (E. Muljadi et al, 1989).

Table 1: Specific parameters of the induction motor under test

Description of data	Values
Main Winding Rated Voltage	380 V
Auxiliary Winding Rated Voltage	380V
Number of poles	4
Magnetizing Reactance (X_m)	37.86 Ω
Main winding phase resistance (R_1)	4.33 Ω

Auxiliary winding phase resistance (R_2)	18.1 Ω
Main winding leakage reactance (X_{l1})	6.97 Ω
Auxiliary winding leakage reactance(X_{l2})	6.97 Ω
Rotor resistance (R_r)	1.35 Ω
Rotor leakage reactance (X_{lr})	1.97 Ω
Full load main winding current	8.6 A

Firstly, the motor was tested without compensation; the results are presented in Table 2.

Table 2: Experimental data for the motor under test

I_L(A)	T_L(Nm)	P(W)	S(VA)	PF	Slip	RPM
5.43	1.83	678.6	3568.7	0.17	0.0026	1496
5.56	4.73	1734	3655.1	0.43	0.0060	1491
6.02	5.49	1992	3957.5	0.49	0.0086	1487
6.88	7.53	2709	4522.9	0.61	0.0153	1477
7.74	9.89	3540	5088.2	0.70	0.0213	1468
8.61	12.1	4179	5653.6	0.74	0.056	1416

Simulation Results

Based on Matlab platform, a simulation model has been built (Figure 11). The parameters from Table 1 have been used for Matlab model after delta-star transformation.

The rotor resistor was simulated using a variable resistor depending on slip. The model was run firstly to show the capability of adjusting the power factor via the static switch.

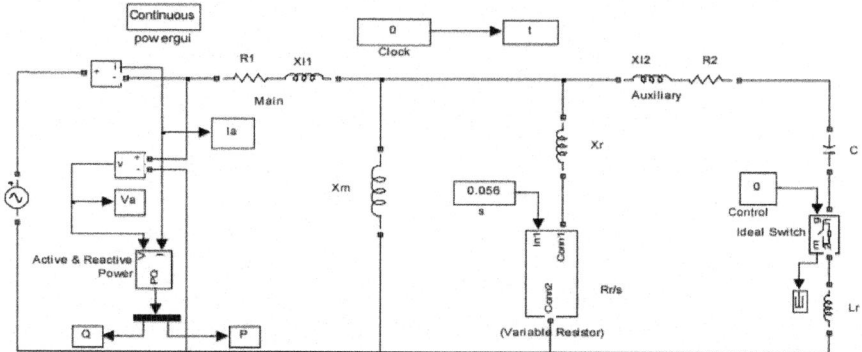

Figure 11: Basic simulation model

Thyristor-based variable capacitor

One of the first solutions to introduce a variable capacitor was to use an "ac-ideal" switch based on two thyristors anti-parallel connected. The test was done using a 100 µF and a delay angle of 45° for a slip of 0.0026 (no load). This results in a reduction of reactive power drawn by the motor from 1193 VAr to 46 VAr. The main drawback of this switching solution is the strong distorting currents introduced as could be seen in Figure 12.

As could be noticed, both currents main and auxiliary are very much distorted. Obviously, the power factor compensation is accompanied with degrading of all other performances of the motor.

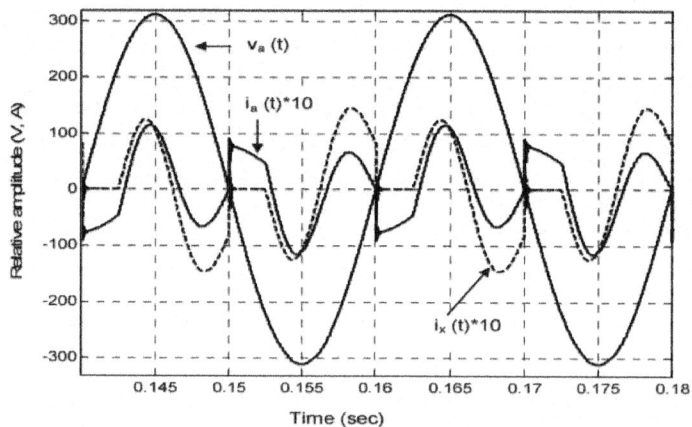

Figure 12: Electric parameters (v_a, i_a and i_x) for compensated model using thyristor-based variable capacitor

IGBT-based variable capacitor

The other solution tested to obtaining variable capacitor was that of Figure 6. The values for the two capacitors were chosen as: $C_1 = 100$ μF and $C_2 = 1$ μF. The validation simulation was done for a slip of $s = 0.0026$; the switching frequency was 1 kHz and then the duty cycle manually adjusted. Figure 13 shows the main and auxiliary currents toward supply voltage for a duty cycle of 40 % (Fig.13. a) and 80% (Fig.13. a).

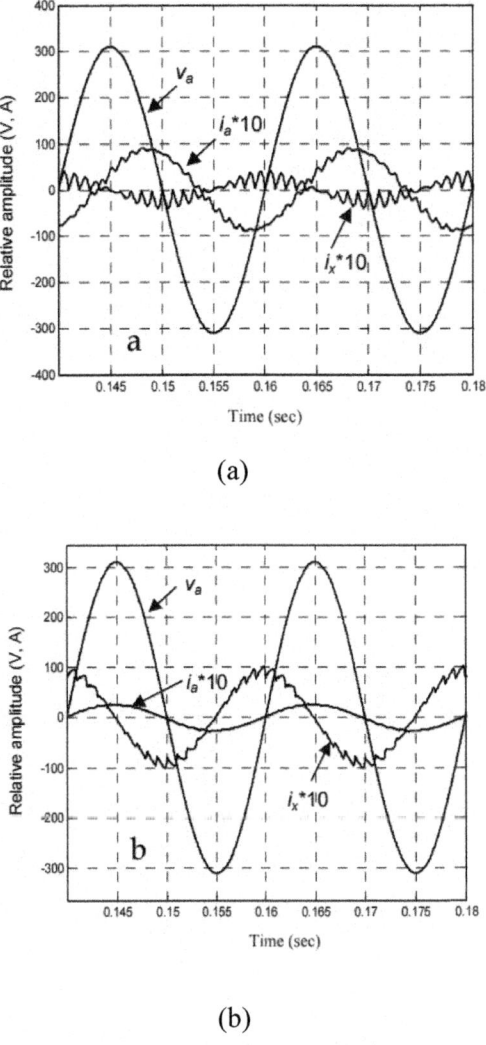

(a)

(b)

Figure 13: Main and auxiliary currents for s = 0.0026, 1 kHz and: a) = 40 %; b) = 80 %.

As can be noticed a duty cycle of 40 % does adjust the shift between the main current and voltage marginally increasing the power factor from 0.15 to 0.32 lagging. When the duty cycle was increased to 80%, then the main current arrived in phase with the voltage. It can also be noticed a relative high ripple especially for low duty cycle.

Figure 14 shows the same simulation conditions, but now the switching frequency was raised to 4 kHz. As can be noticed the ripple has been reduced significant. The other methods of producing variable capacitive reactive power, meaning the H-bridge topologies require a more complex control system and are not addressed in this study.

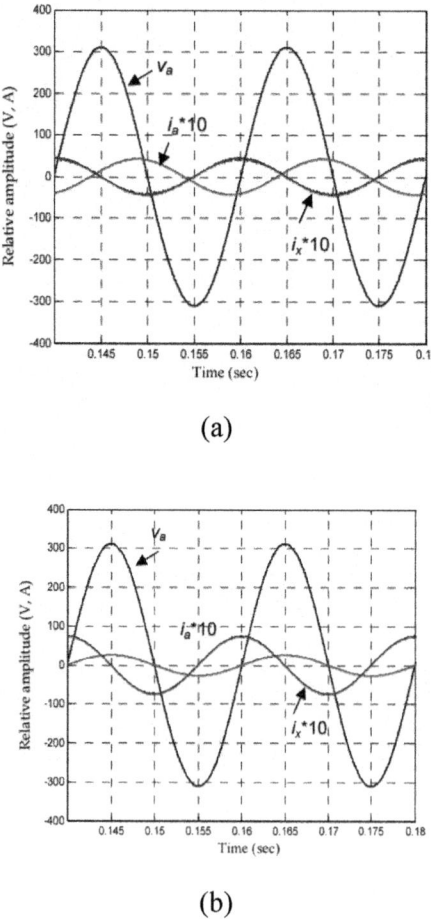

Figure 14: Main and auxiliary currents for s = 0.0026, 4 kHz and: a) = 40 %; b) = 80 %.

Capacitance versus slip for unity power factor

Now, the model was run for each value of the slip (speed) given in Table 2. The result of these simulations is the capacitance producing unity power factor (see Table 3).

Table 3: Simulation results: capacitance versus slip for unity power factor

s	CY	Ia	Pa	Q_a	S_a	PF
(n)	(µF)	(A)	(W)	(VAr)	(VA)	
0.0026	0	5.49	184	1193	1207	0.152
(1496)	81	1.65	362	9	363	0.999
0.006	0	5.65	366	1187	1242	0.295
(1491)	81	2.57	565	11	565	0.999
0.0086	0	5.86	505	1186	1289	0.392
(1487)	81.6	3.27	718	15	719	0.999
0.0153	0	6.71	855	1200	1473	0.58
(1477)	87	5.21	1143	1	1143	0.999
0.0213	0	7.68	1160	1229	1689	0.686
(1468)	90	6.82	1499	17	1499	0.999
0.056	0	8.31	1330	1252	1826	0.728
(1491)	90.6	7.66	1684	27	1686	0.999
0.0026	75	1.57	330	101	345	0.956
0.006	75	2.47	531	111	542	0.978
0.0086	75	3.16	683	123	694	0.984
0.0153	75	4.91	1066	170	1079	0.987
0.0213	75	6.44	1396	230	1415	0.987
0.056	75	7.29	1580	272	1603	0.986

It is interesting to observe that the capacitance producing unity power factor does not vary so much as presented in other studies (E. Muljadi et al. 1989; Tamrakan and Malik, 1999) concerning similar method of injecting capacitive reactive power.

Moreover, it could be found a fix value of 75 µF which maintain a high power factor irrespective of the load/slip. Table 4 shows the simulation results for the fixed capacitor and variable load/slip. Figure 15 shows the variation of the power factor versus slip and Figure 16 shows the variation of the supply current versus slip for compensated with a fix 75 µF capacitor per phase and uncompensated machine.

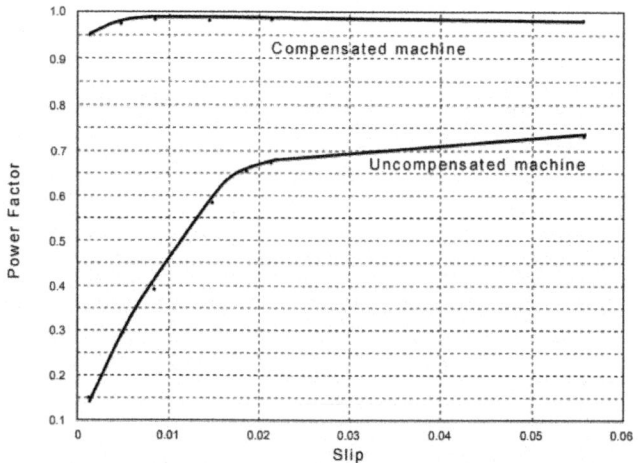

Figure 15: Power Factor versus slip

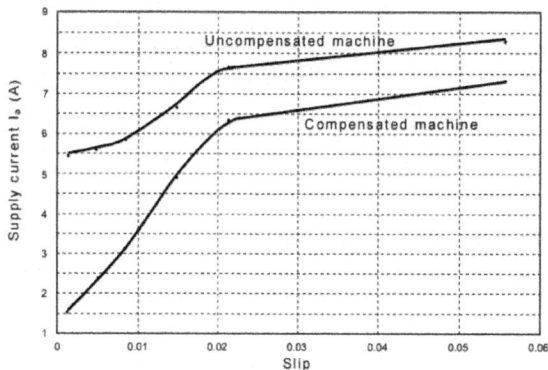

Figure 16: Supply current versus slip

Experimental Validation

After rewinding the machine under test with both sets of windings delta connected, three identical capacitors were connected to the auxiliary. Then the "compensated" motor was loaded gradually to get the same speeds as for the "uncompensated" motor. The testing started with a capacitance of 25 µF as found during the simulation: $C = C_Y / 3$. This did not produce the results found via the simulation. Then a value of 28 µF was found to give a relative even power factor of approximate 0.95 in the entire loading range (see Table 4).

Figure 17 shows the phase voltage (v_{an}) and current (i_a) for the uncompensated (Fig. 17.a) and compensated (Fig. 17.b) with slip of 0.0026. It can be observed, the power factor increased from 0.17 to 0.937 which represent 450% improvement for no load. For full load the improvement in power factor is 29%.

Table 2: Experimental data for the motor under test

I_L(A)	T_L(Nm)	P(W)	S(VA)	PF	Slip	RPM
1.71	2.81	1058	1129	0.937	0.0026	1496
2.71	4.73	1694	1789	0.947	0.0060	1491
3.49	5.83	2191	2304	0.951	0.0086	1487
5.46	7.63	2709	3604	0.955	0.0153	1477
6.92	9.29	3442	4567	0.952	0.0213	1468
7.85	13.8	4937	5181	0.953	0.056	1416

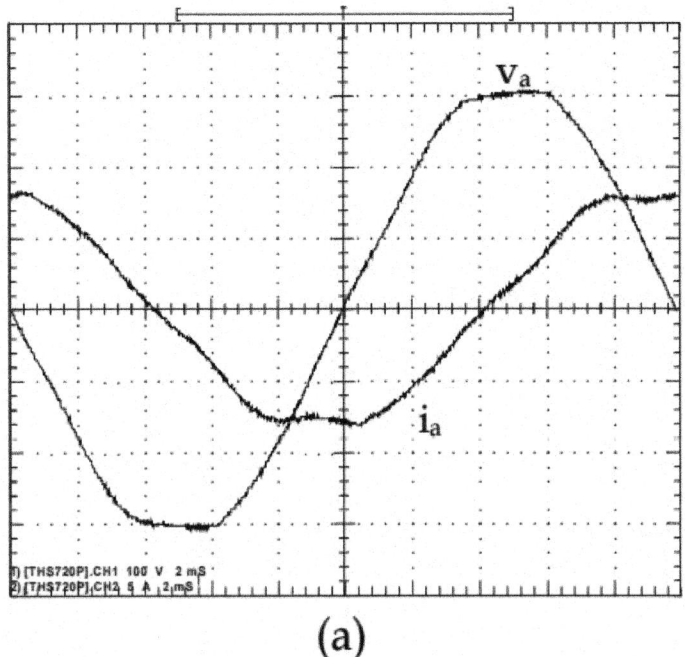

(a)

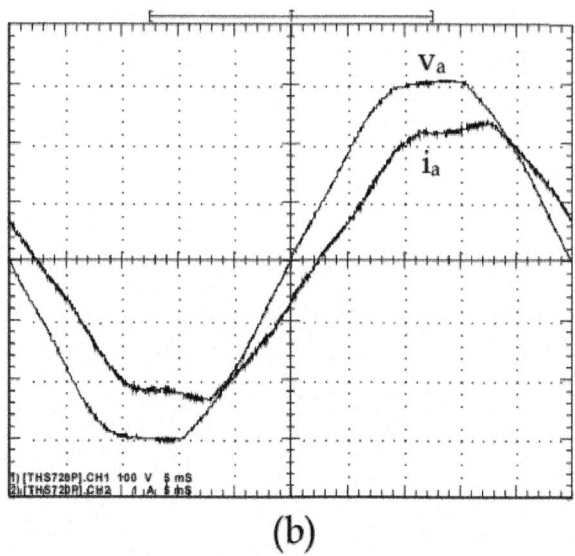

Figure 17: Voltage and current for: a) uncompensated motor; b) compensated motor

COMMENTS

This study proved that direct injecting capacitance reactance through auxiliary winding does improve the power factor of the induction motor; it also increases the ratio torque over current. What is also very interesting is this method achieves a "flat" variation of power factor with respect of load variation. This is a very important improvement given the fact that majority of induction motors do not work at constant full load where the classic design produces maximum performances.

However, this method introduces extra copper losses reducing the overall efficiency and increasing the operation temperature.

This study did not intended to elucidate the full effect of the method upon the torque especially the existence of the influence produced by the current flowing through auxiliary windings. The only aspect clearly noticed was the torque ripple introduced by the "single-phase" auxiliary winding connection.

Further more, the same method was applied to a synchronous reluctance motor (Ogunjuyigbe *et al*, 2010). The same type of stator winded similarly as above was used with a salient milled rotor obtained from the corresponding squirrel cage induction motor. The experimental results show an improvement of the power factor from 0.41-0.69 to 0.93-0.97 for the entire loading range. The economic benefits are related with the savings on demand especially for places where a large number of three-phase induction motors are used under variable loading.

REFERENCES

1. M. A El-Sharkawi, S. S Venkata, T. J Williams, and N. G Butler, "An adaptive Power Factor Controller for Three-Phase Induction Generators", Paper 84 SM 672-2 presented at the *IEEE/PES Summer Meeting,* Seattle, Washington, July 15-20, 1984.
2. Fuchs, E.F. Hanna, W.J.; "Measured efficiency improvements of induction motors with thyristor/triac controllers", , IEEE Transaction on Energy Conversion, Volume 17, Issue 4, Dec. 2002 pp. 437 – 444
3. *Suciu, C.; Dafinca, L.; Kansara, M.; Margineanu, I.; "Switched capacitor fuzzy control for power factor correction in inductive circuits",* IEEE 31st Annual Power Electronics Specialists Conference, 2000. *Vol. 2, pp. 773 - 777*
4. C. Suciu, M. Kansara, P. Holmes and W. Szabo, "Performance Enhancement of an Induction Motor by Secondary Impedance Control, *IEEE Trans. On Energy Conversion,* Vol. 17, No. 2, June 2002
5. J. Reinert, M.J. Parsley, "Controlling the speed of 8×1 induction motor by resonating the rotor &wit," in *IEEE Transactions on Industry Applications,* Vol. 31, No. 4, July/August 1995, pp. 887-891.
6. E. Muljadi, T.A. Lipo, D.W. Novotny, "Power Factor Enhancement of Induction Machines by Means of Solid State Excitation," *IEEE Trans. on Power Electronics,* Vol. 4, No. 4, pp. 409418, Oct. 1989.
7. I.M Tamrakan and O.P Malik, "Power Factor Correction of Induction motors Using PWM Inverter Fed Auxiliary Stator Winding", *IEEE Transaction on Energy Conversion,* Vol. 14, No.3, Sept, 1999, pp. 426-432
8. I B. Medarametla, M. D. Cos, and Baghzouz, "Calculations and measurement of unity plus three-phase induction motor," *IEEE Transactions on Energy Conversion,* vol. 7, no. 4, pp. 732-738, 1992.
9. S.D. Umans, and H. L. Hess, "Modelling and analysis of a the Wanlass three-phase induction motor configuration," *IEEE Transaction on Power Apparatus and Systems,* vol. 102, no. 9, pp. 2912-2916, 1983.
10. R. Spée and A. Wanllace, "Comparative Evaluation Of Power-Factor Improvement Techniques For Squirrel cage Induction Motors", Industry

Applications Society Annual Meeting, 1990.

11. A.A. Jimoh and D.V. Nicolae, "Performance Analysis of a Three-Phase Induction Motor with Capacitance Injection", OPTIM'06, 10th International Conference on Optimization of Electrical and Electronic Equipments, Brasov, Romania, May 17-20, 2006

12. D.V. Nicolae and A.A. Jimoh, "Three-Phase Induction Motor with Power Electronic Controlled Single-Phase Auxiliary Stator Winding", PESC'07, The 38th IEEE Power Electronics Specialists Conference, Orlando, USA, June 17-21, 2007

13. A.S.O. Ogunjuyigbe, A.A. Jimoh, D.V. Nicolae and E.S. Obe, "Analysis of Synchronous Reluctance Machine with Magnetically Coupled Three Phase Windings and Reactive Power compensation", *IET Electric Power Applications,* 2010, Vol. 4, Iss. 4, pp 291-303

Chapter 5

MAGNETIC RELUCTANCE METHOD FOR DYNAMICAL MODELING OF SQUIRREL CAGE INDUCTION MACHINES

Jalal Nazarzadeh and Vahid Naeini

Faculty of Engineering Shahed University, Tehran Iran

INTRODUCTION

Nowadays, induction machines play important role in electromechanical energy conversion in industry. These machines are often operated in critical conditions where can cause unexpected failures and outages. Generally, stator and bearing faults, broken rotor bar and end-rings, air-gap irregularities are some of the major faults in an induction machine (Al-Shahrani, 2005; Sprooten, 2007) which may be situated the induction machines in out of service (Siddique et al., 2005). Fourier analysis for stator currents (Bellini et al., 2001; Benbouzid, 2000; Jung et al., 2006), torque and rotor speed, acoustic noise and temperature analysis (Siddique et al., 2005) are some classical techniques which introduced for identification and diagnosis of induction machines faults. Additionally, other heuristic methods were proposed to monitor of the induction machines for fault detection. For instance, neural network modelling were applied to monitor an induction machine for fault detection (Su & Chong, 2007). Also, space vector of rotor magnetic field (Mirafzal & Demerdash, 2004) based on artificial intelligent approaches and pendulous oscillation of the rotor magnetic field were proposed. Recently, a new technique based on the analysis of three-phase stator current envelopes was presented (Mirafzalet & Demerdash, 2008). In all monitoring and fault detection techniques, we need to tune up the monitoring systems based on response of induction machines for proper operations. However, experimental set up for testing any arbitrary fault conditions are not practical. Thus, an accuracy dynamic and steady state models of induction machines are very important for this propose.

Also, for dynamical modelling of induction machines, space harmonic distribution, core saturation and loss are often neglected in abc quantitative and two-axis methods (Krause et al., 1995). Thus, these approaches do not have an efficient accuracy for modelling of induction machines in asymmetrical and non-linear conditions. For considering distribution rotor bars, coupled magnetic circuit method (Muñoz & Lipo, 1999), abc quantitative based on rotor bar currents (Alemi & Nazarzadeh, 1996) can be utilized. Furthermore, winding function method may be used to include the stator winding distribution effect in the air gap flux (Luos et al., 1995). However, in all mentioned methods, the core saturation, stator and rotor teeth effects and distributions of the rotor and stator windings cannot be investigated, simultaneously. Also, Finite Element Method (FEM) is a professional technique for analysis of any electromagnetic systems, which needs to magnetic and geometry details of the systems (Faiz et al., 2002). This method is very accurate and flexible, but due to complexity, the dynamical modelling of an induction machine is quite complicated. Contrary to FEM, the Magnetic Equivalent Circuit Method (MECM) can apply to analysis of the electro magnetics problems with lower complexity. Magnetic saturation, space harmonics in stator and rotor teeth, stator windings and distributed structure of a squirrel-cage rotor can be considered by MECM for modelling and analysis of any induction machines (Jeong et al., 2003; Ostovic, 1989). In this approach, the non-linear reluctances of flux paths use to configure magnetic equivalent circuit. This method has less complexity than FEM for dynamical modelling of induction machines. Therefore, developing an exact details model of induction machine for analysis of transient, sensitivity and fault diagnosis in the asymmetrical conditions are very essential.

The present chapter introduces methodology of MECM for modelling and analysis of asymmetrical non-linear systems in transient and steady state conditions. MECM is very suitable method for finding a generalized accurate dynamical model of squirrel cage induction machines with asymmetrical conditions. For evaluation of the method, several simulations in linear and non-linear conditions are made. Also, some simulations results for induction machines with broken bar faults and core saturation conditions are included to illustrate capability of the method in asymmetrical conditions.

ELECTRIC AND MAGNETIC BASED MODEL OF SQUIRREL CAGE INDUCTION MACHINES

For detailed modelling of any electromagnetic systems, we have to find a correlation between electric and magnetic variables of the system. Generally, a set of non-linear differential equations presents dynamical model of a electromagnetic system that by using numerical analysis, transient response of the electrical variables can be obtained. In addition, non-linear algebraic equations illustrate non-linear relations between electrical and magnetic variables. MECM provides an augmented model of the electromagnetic systems, in which we can achieve all variables of the systems in transient and steady state, simultaneously. Also, the main advantages of MECM for modelling of induction machines are; simple algorithm for including distribution winding, stator and rotor teeth effects and magnetic core saturation phenomena. Global non-linear model of squirrel cage induction machines can be offered in algebraical (magnetic) and differential (electric) equations which will be presented in the following sections.

MAGNETIC EQUIVALENT CIRCUIT OF INDUCTION MACHINES

Fig. 1 shows a part of rotor and stator structures for a typical squirrel cage induction machine that magnetic circuit elements are presented for rotor and stator teeth and yoke. Numbers of rotor and stator teeth are considered by h and l, respectively. Also in this figure, magnetic mutual permeances of rotor and stator teeth in air gap are shown. Due to non-linear characteristics of flux and magnetic current in iron core, permeances of the rotor and stator in the magnetic cores are illustrated as non-linear elements. $G_{i,j}$ is linear permeance of flux path between i^{th} stator and j^{th} rotor teeth in the air gap.

Magnetic Node Equations

Due to the fact that magnetic permeances of each stator and rotor teeth make several magnetic loops in the air gap, we apply node magnetic potential equations to the each air gap nodes for simplicity in algebraic magnetic equations. For instance, sum of magnetic fluxes in i^{th} air gap node (stator tooth) have to be zero; thus we can write:

$$(u_i^s - u_{i+1}^s) G^{s\sigma} + (u_i^s - u_{i-1}^s) G^{s\sigma} + (u_i^s - u_1^r) G_{i1} + \cdots + (u_i^s - u_l^r) G_{il} = \phi_i^{st} \quad (1)$$

where u_i^s, u_i^r and ϕ_i^{st} are magnetic potential of stator, rotor and flux of ith stator tooth and $G^{s\sigma}$ and $G_{i,j}$ are permeances of stator neighbour slots and mutual permance between i^{th} stator with j^{th} rotor teeth, respectively.

Similar Eq. (1), continuity principle in the teeth fluxes at i^{th} rotor node yields as:

$$(u_i^r - u_{i+1}^r) G^{r\sigma} + (u_i^r - u_{i-1}^r) G^{r\sigma} + (u_i^r - u_1^s) G_{i1} + \cdots + (u_i^r - u_h^s) G_{ih} = -\phi_i^{rt} \quad (2)$$

In Eq. (2), ϕ_i^{st} and $G^{r\sigma}$ are flux of ith rotor tooth and permeance of rotor slot, respectively. Consequently, node potential equations for equivalent circuit of Fig. 1 along the air gap can be written as:

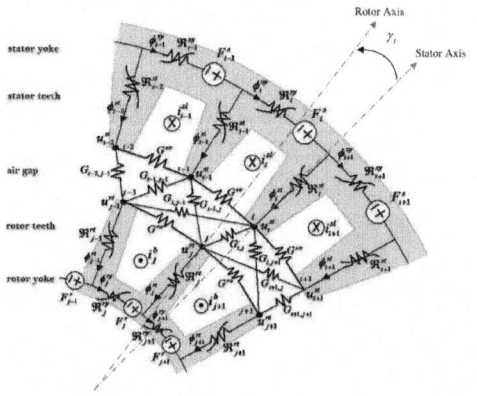

Figure 1: Magnetic equivalent circuit of induction machine

$$\mathbf{A}_{ss}\mathbf{U}_{st} + \mathbf{A}_{sr}\mathbf{U}_{rt} = \mathbf{\Psi}_{st} \qquad (3)$$

$$\mathbf{A}_{rs}\mathbf{U}_{ss} + \mathbf{A}_{rr}\mathbf{U}_{rt} = -\mathbf{\Psi}_{rt} \qquad (4)$$

where $\mathbf{A}_{ss} \in \mathbb{R}^{h \times h}$, $\mathbf{A}_{rr} \in \mathbb{R}^{l \times l}$ and $\mathbf{A}_{sr} \in \mathbb{R}^{h \times l}$ are air gap permeance coefficients matrices that we can written as:

$$\mathbf{A}_{ss} = \begin{bmatrix} 2G^{s\sigma}+\sum_{j=1}^{l} G_{1,j} & -G^{s\sigma} & 0 & \cdots & 0 & 0 & -G^{s\sigma} \\ -G^{s\sigma} & 2G^{s\sigma}+\sum_{j=1}^{l} G_{2,j} & -G^{s\sigma} & \cdots & 0 & 0 & 0 \\ \vdots & \vdots & \vdots & \ddots & \vdots & \vdots & \vdots \\ -G^{s\sigma} & 0 & 0 & \cdots & 0 & -G^{s\sigma} & 2G^{s\sigma}+\sum_{j=1}^{l} G_{h,j} \end{bmatrix} \qquad (5)$$

$$\mathbf{A}_{rr} = \begin{bmatrix} 2G^{r\sigma}+\sum_{i=1}^{h} G_{i,1} & -G^{r\sigma} & 0 & \cdots & 0 & 0 & -G^{r\sigma} \\ -G^{r\sigma} & 2G^{r\sigma}+\sum_{i=1}^{h} G_{i,2} & -G^{r\sigma} & \cdots & 0 & 0 & 0 \\ \vdots & \vdots & \vdots & \ddots & \vdots & \vdots & \vdots \\ -G^{r\sigma} & 0 & 0 & \cdots & 0 & -G^{r\sigma} & 2G^{r\sigma}+\sum_{i=1}^{h} G_{i,l} \end{bmatrix} \qquad (6)$$

$$\mathbf{A}_{sr} = \mathbf{A}_{rs}^{T} = \begin{bmatrix} G_{1,1} & G_{1,2} & \cdots & G_{1,l} \\ G_{2,1} & G_{2,2} & \cdots & G_{2,l} \\ \vdots & \vdots & \ddots & \vdots \\ G_{h,1} & G_{h,2} & \cdots & G_{h,l} \end{bmatrix} \qquad (7)$$

Also, $\mathbf{\Psi}_{st} \in \mathbb{R}^{h \times 1}$, $\mathbf{\Psi}_{rt} \in \mathbb{R}^{l \times 1}$, $\mathbf{U}_{st} \in \mathbb{R}^{h \times 1}$ and $\mathbf{U}_{rt} \in \mathbb{R}^{l \times 1}$ are stator and rotor teeth fluxes vectors, stator and rotor magnetic scalar potentials vectors, respectively. These vectors can be presented as:

$$\Psi_{st} = \begin{bmatrix} \phi_1^{st} & \phi_2^{st} & \cdots & \phi_h^{st} \end{bmatrix}^T \tag{8}$$

$$\Psi_{rt} = \begin{bmatrix} \phi_1^{rt} & \phi_2^{rt} & \cdots & \phi_l^{rt} \end{bmatrix}^T \tag{9}$$

$$U_{st} = \begin{bmatrix} u_1^{st} & u_2^{st} & \cdots & u_h^{st} \end{bmatrix}^T \tag{10}$$

$$U_{rt} = \begin{bmatrix} u_1^{rt} & u_2^{rt} & \cdots & u_l^{rt} \end{bmatrix}^T \tag{11}$$

In Eq. (5), $G^{s\sigma}$ is linear permeances between each successive stator slots with constant geometric permeability. This permeance can be obtained from (see Fig. 1):

$$G^{s\sigma} = \mu_0 \frac{A^{s\sigma}}{L^{s\sigma}} \tag{12}$$

Similarly, for $G^{r\sigma}$ we can write:

$$G^{r\sigma} = \mu_0 \frac{A^{r\sigma}}{L^{r\sigma}} \tag{13}$$

where $A^{s\sigma}$, $L^{s\sigma}$, $A^{r\sigma}$, $L^{r\sigma}$ and μ_0 are cross section and length of stator and rotor slot opening and air permeability, respectively. Furthermore, mutual permeance between i^{th} and j^{th} slots of the rotor and stator teeth depends to rotor mechanical angle ($\gamma_i(t)$), momentarily. Fig.2-a shows a typical geometry of the stator and rotor teeth. In this case, stator and rotor mutual permeance can be approximated by:

$$G_{i,j}(\gamma) = \begin{cases} G_{max} & 0 < \gamma < \gamma_t'\quad 2\pi - \gamma_t' < \gamma < 2\pi \\ G_{max}\dfrac{1+\cos\pi\frac{\gamma-\gamma_t'}{\gamma_t-\gamma_t'}}{2} & \gamma_t' < \gamma < \gamma_t \\ G_{max}\dfrac{1+\cos\pi\frac{\gamma-2\pi+\gamma_t'}{\gamma_t-\gamma_t'}}{2} & 2\pi-\gamma_t < \gamma < 2\pi-\gamma' \\ 0 & \gamma_t < \gamma < 2\pi-\gamma_t \end{cases} \tag{14}$$

where γ_t and γ'_t are two mechanical angles which are depended to the stator and rotor teeth geometry. For instance, in Fig. 2-a; γ_t and γ'_t can be obtained as:

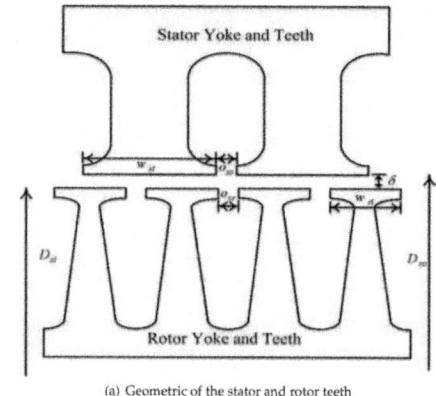

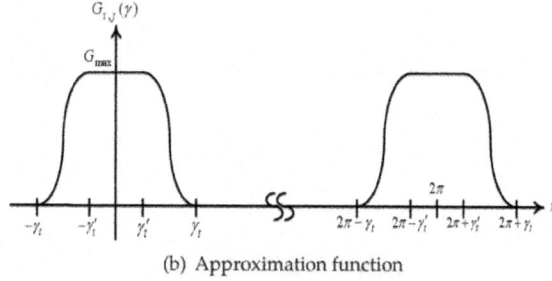

(b) Approximation function

Figure 2: A typical air gap permeance between stator and rotor teeth

$$\gamma_t = \frac{w_{st} + w_{rt} + o_{ss} + o_{sr}}{D_{ag}} \quad (15)$$

$$\gamma_t' = \frac{|w_{st} - w_{rt}|}{D_{ag}} \quad (16)$$

where w_{st}, w_{rt}, o_{ss} and o_{sr} are dimensions of stator and rotor which are shown in Fig. 2-a and D_{ag} is :

$$D_{ag} = \frac{D_{si} + D_{ro}}{2} \quad (17)$$

The maximum value of air gap permeance G_{max} can be written as

$$G_{max} = \mu_0 \frac{l \times \min[w_{st}, w_{rt}]}{\delta} \quad (18)$$

where l and δ are lengths of machine and air gap.

Magnetic Mesh Equations

Eqs. (3) and (4) present l + h node equations which are shown relation between magnetic potentials of teeth nodes with stator and rotor teeth fluxes. Attention to Fig. 1, two neighbour stator teeth and yoke paths make a simple mesh in each stator slots, that sum of the magnetic potentials in these mesh have to be zero. For instance, magnetic mesh equation in i^{th} and $(i + 1)^{th}$ stator teeth and yoke can be written as:

$$u_i^{st} - u_{i-1}^{st} - \Re_{i-1}^{st}\phi_{i-1}^{st} + \Re_i^{st}\phi_i^{st} + \Re_i^{sy}\phi_i^{sy} = F_i^S \tag{19}$$

Similarly, mesh equation for i^{th} rotor tooth and yoke with $(i + 1)^{th}$ rotor tooth can be obtained as:

$$u_i^{rt} - u_{i-1}^{rt} + \Re_{i-1}^{rt}\phi_{i-1}^{rt} - \Re_i^{rt}\phi_i^{rt} + \Re_i^{ry}\phi_i^{ry} = F_i^r \tag{20}$$

Thus, magnetic mesh equations for all rotor and stator meshes can be expressed as:

$$\mathbf{A}_{ust}\mathbf{U}_{st} + \mathbf{A}_{\Psi_{sy}}\mathbf{\Psi}_{sy} + \mathbf{A}_{\varphi st}\mathbf{\Psi}_{st} = \mathbf{F}_s \tag{21}$$

$$\mathbf{A}_{urt}\mathbf{U}_{rt} + \mathbf{A}_{\varphi ry}\mathbf{\Psi}_{ry} + \mathbf{A}_{\varphi rt}\mathbf{\Psi}_{rt} = \mathbf{F}_r \tag{22}$$

In Eqs. (21) and (22), $\mathbf{A}_{\varphi st} \in \mathbb{R}^{h \times h}$ and $\mathbf{A}_{\varphi rt} \in \mathbb{R}^{l \times l}$ are diagonal coefficients matrices of the stator and rotor teeth reluctance that we can find as:

$$\mathbf{A}_{\varphi st} = \begin{bmatrix} \Re(B_1^{st}) & 0 & 0 & \cdots & 0 & -\Re(B_h^{st}) \\ -\Re(B_1^{st}) & \Re(B_2^{st}) & 0 & \cdots & 0 & 0 \\ 0 & -\Re(B_2^{st}) & \Re(B_3^{st}) & \cdots & 0 & 0 \\ \vdots & \vdots & \vdots & \ddots & \vdots & \vdots \\ 0 & \cdots & 0 & \cdots & \Re(B_{h-1}^{st}) & 0 \\ 0 & \cdots & 0 & \cdots & -\Re(B_{h-1}^{st}) & \Re(B_h^{st}) \end{bmatrix} \tag{23}$$

$$\mathbf{A}_{\varphi rt} = \begin{bmatrix} -\Re(B_1^{rt}) & 0 & 0 & \cdots & 0 & \Re(B_l^{rt}) \\ \Re(B_1^{rt}) & -\Re(B_2^{rt}) & 0 & \cdots & 0 & 0 \\ 0 & \Re(B_2^{rt}) & -\Re(B_3^{rt}) & \cdots & 0 & 0 \\ \vdots & \vdots & \vdots & \ddots & \vdots & \vdots \\ 0 & \cdots & 0 & \cdots & -\Re(B_{l-1}^{rt}) & 0 \\ 0 & \cdots & 0 & \cdots & \Re(B_{l-1}^{rt}) & -\Re(B_l^{rt}) \end{bmatrix} \tag{24}$$

also $A_{\varphi sy} \in R^{h\times h}$ and $A_{\varphi ry} \in R^{l \times l}$ are stator and rotor yoke reluctance coefficients matrices which can be written as

$$A_{\varphi sy} = diag\left(\Re(B_1^{sy}), \Re(B_2^{sy}), \cdots, \Re(B_h^{sy})\right) \tag{25}$$

$$A_{\varphi ry} = diag\left(\Re(B_1^{ry}), \Re(B_2^{ry}), \cdots, \Re(B_l^{ry})\right) \tag{26}$$

which $\Psi_{sy} \in R^{h\times 1}$, $\Psi_{ry} \in R^{l \times 1}$ are stator and rotor yoke fluxes vectors and $F_s \in R^{h\times 1}$ and $F_r \in R^{l \times 1}$ are stator and rotor ampere-turn vectors, respectively. These vectors are considered as:

$$\Psi_{sy} = \begin{bmatrix} \phi_1^{sy} & \phi_2^{sy} & \cdots & \phi_h^{sy} \end{bmatrix}^T \tag{27}$$

$$\Psi_{ry} = \begin{bmatrix} \phi_1^{ry} & \phi_2^{ry} & \cdots & \phi_l^{ry} \end{bmatrix}^T \tag{28}$$

$$F_s = \begin{bmatrix} F_1^s & F_2^s & \cdots & F_h^s \end{bmatrix}^T \tag{29}$$

$$F_r = \begin{bmatrix} F_1^r & F_2^r & \cdots & F_l^r \end{bmatrix}^T \tag{30}$$

Also, $A_{ust} \in R^{h\times h}$ and $A_{urt} \in R^{l \times l}$ are constant matrices which are given by:

$$A_{ust} = A_{urt} = \begin{bmatrix} 1 & 0 & 0 & \cdots & 0 & -1 \\ -1 & 1 & 0 & \cdots & 0 & 0 \\ 0 & -1 & 1 & \cdots & 0 & 0 \\ \vdots & \vdots & \vdots & \ddots & \vdots & \vdots \\ 0 & 0 & 0 & \cdots & 1 & 0 \\ 0 & 0 & 0 & \cdots & -1 & 1 \end{bmatrix} \tag{31}$$

Teeth and yoke Flux Relations

For completing of the magnetic model of an induction machine, we need to find a relation between teeth and yoke fluxes in mesh equations (Eqs. (21) and (22)). Attention to Fig. 1, the relations between stator teeth and yoke fluxes can be obtained by applying magnetic flux continuity principle in the yoke nodes. Thus, we have:

$$\phi_i^{sy} = \phi_i^{st} + \phi_{i+1}^{sy} \tag{32}$$

Similarly, relation between the rotor fluxes are:

$$\phi_{j+1}^{ry} = \phi_j^{rt} + \phi_j^{ry} \tag{33}$$

Eqs. (32) and (33) can be presented in matrix form as:

$$\Psi_{st} = A_{syt}\Psi_{sy} \tag{34}$$

$$\Psi_{rt} = A_{ryt}\Psi_{ry} \tag{35}$$

where $A_{syt} \in R^{h \times h}$ and $A_{ryt} \in R^{l \times l}$ can be written as

$$A_{ryt} = -A_{syt} = \begin{bmatrix} -1 & 1 & 0 & \cdots & 0 & 0 \\ 0 & -1 & 1 & \cdots & 0 & 0 \\ 0 & 0 & -1 & \cdots & 0 & 0 \\ \vdots & \vdots & \vdots & \ddots & \vdots & \vdots \\ 0 & 0 & 0 & \cdots & -1 & 1 \\ 1 & 0 & 0 & \cdots & 0 & -1 \end{bmatrix} \tag{36}$$

In a squirrel cage induction machine, a stator winding is not concentrated in single slot, but it is distributed along air gap for harmonics reduction, full utilization of core and reduction of mechanical stress to the winding. Thus, flux of stator coil equals the sum of fluxes of stator teeth in the coil pitch. If three phase flux vector of the stator windings denotes as $\Psi_{3\varphi}$, we can find a matrix relation between stator teeth and windings fluxes as:

$$\Psi_{3\phi} = M_{cf}\Psi_{st} \tag{37}$$

where $M_{cf} \in R^{3 \times h}$ is a connected matrix which can be obtained based on the connection diagram of stator windings.

CORE SATURATION CHARACTERISTIC

Generally, magnetic cores of any electrical machines have non-linear characteristic curve (B − H), thus the elements of reluctance matrices in Eqs. (23) and (24) are depended to their fluxes. For inserting the non-linearity characteristic of the magnetic core to mesh equations, a non-linear permeability dependent to the core field density are used. For this purpose, a non-linear permeability is defined as:

$$\mu(B) = \mu_0 \mu_r(B) = \frac{\partial B}{\partial H} \tag{38}$$

So, the reluctances of flux path in the stator and rotor cores can be written as:

$$\Re\left(B_i^k\right) = \frac{L_i^k}{\mu(B_i^k)A_i^k} \quad \begin{array}{l} k = sy, st, ry, rt \\ i = 1, 2, \ldots \end{array} \tag{39}$$

where L_i^k and A_i^k are i^{th} length and cross section of flux path in the stator and rotor cores, respectively. Tangent, exponential and piecewise linear functions may be used to approximate the saturation curve (Chen et al., 2005). In this chapter, the magnetic core permeability is considered as:

$$\mu(B) = b \times e^{a \times B^2} \tag{40}$$

where a and b are two constants which can be chosen for the best fitting of the saturation curve (B − H) with this relation. By integrating Eq. (38) and combining with Eq. (40), the magnetic field intensity H can be written as

$$H(B) = \int \frac{1}{\mu(B)} dB = \frac{\sqrt{\pi} Erf(\sqrt{a} \times B)}{2b\sqrt{a}} \tag{41}$$

where Er f is error function which is defined as: (Gautschi, 1964):

$$Erf(B) = \int e^{B^2} dB \tag{42}$$

For instance, Fig. 3 shows a saturation curve (B − H) for a silicon steel core that we can find two constants in Eq. (40) for the best fitting of the saturation curve. In this case, the two constants can be found as:

$$a = -0.8$$
$$b = 1000 \tag{43}$$

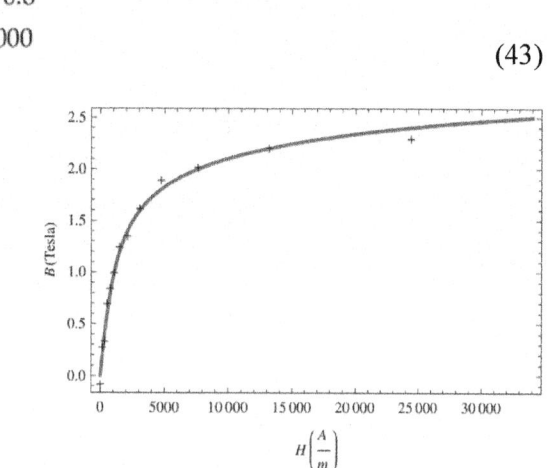

Figure 3: Core saturation curve

EXCITATION VECTORS OF RELUCTANCE NETWORK

For determining relations between three phase currents with ampere-turn in the stator magnetic circuits, we give a contour between $(i-1)^{th}$ and i^{th} stator teeth, yoke and slot in Fig. 1. Also, we suppose that combining of three phase conductors are placed in the stator slots. Therefore, by applying Ampere's law to this contour, $(i)^{th}$ ampere-turn of the stator magnetic circuits (F_i^s) in Eq. (21) can be expressed as:

$$F_i^s = w_i^a i_a + w_i^b i_b + w_i^c i_c \qquad (44)$$

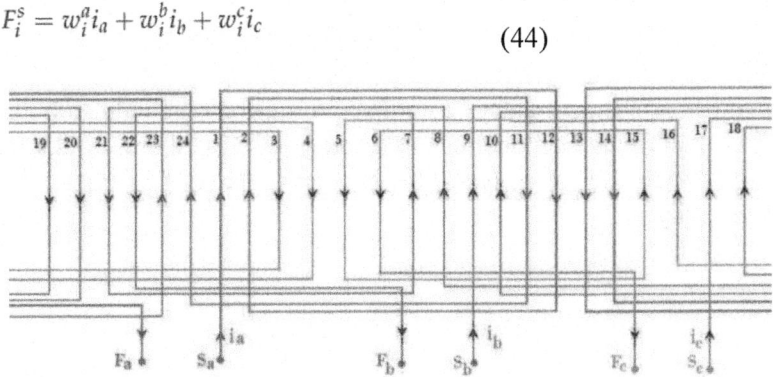

Figure 4: Single layer distributed winding with 24 slot

where w_i^a, w_i^b and w_i^c are conductor numbers of phase a, b and c in the ith stator slot, respectively. Values and signs of these parameters will depend to connections and directions of windings in the ith slot. For instance, Fig. 4 shows a single layer three phase winding with two poles and 24 slots, which can generate a rotating magnetic field in the air gap. In this case, relation between ampere-turn of the stator magnetic circuits and stator currents can be determined as:

$$\begin{bmatrix} F_1^s \\ F_2^s \\ F_3^s \\ \vdots \\ F_{22}^s \\ F_{23}^s \\ F_{24}^s \end{bmatrix} = \begin{bmatrix} -w^a & 0 & 0 \\ -w^a & 0 & 0 \\ 0 & 0 & w^c \\ \vdots & \vdots & \vdots \\ 0 & w^b & 0 \\ -w^a & 0 & 0 \\ -w^a & 0 & 0 \end{bmatrix} \begin{bmatrix} i_a \\ i_b \\ i_c \end{bmatrix} \qquad (45)$$

Thus, the stator ampere-turn vector (F_s) in Eq. (21) may be presented as

$$\mathbf{F}_s = \mathbf{W}_c \mathbf{i}_s \qquad (46)$$

where $W_c \in R^{h \times 3}$, is a connection matrix that shows number of conductors in all of the stator slots. Also, i_s is a vector of three phase currents that we can arrange as:

$$i_s = \begin{bmatrix} i_a & i_b & i_c \end{bmatrix}^T \tag{47}$$

In single layer winding, each slot can carry only one phase current, thus rows of the connection matrix in Eq. (46) have only one non-zero element. These non-zero elements can be determined by attention to stator winding topology.

In other respect, currents of rotor bars in squirrel cage induction machines are equal to current of conductors in the rotor slots. Therefore, similar to above mentioned method; the ampere-turn in rotor slots (F_r) in Eq. (22) can be determined by:

$$F_r = i_b \tag{48}$$

where i_b, the rotor bars currents vector is defined as

$$i_b = \begin{bmatrix} i_1^b & i_2^b & \cdots & i_l^b \end{bmatrix}^T \tag{49}$$

Substituting Eqs. (34), (35), (46) and (48) into (21) and (22) and combining with Eqs. (3) and (4), the magnetic algebraic equations of squirrel cage induction machines in matrix form can be augmented as

$$\begin{bmatrix} A_{\phi sy}+A_{\phi st}A_{syt} & 0 & A_{ust} & 0 \\ 0 & A_{\phi ry}+A_{\phi rt}A_{ryt} & 0 & A_{urt} \\ -A_{syt} & 0 & A_{ss} & A_{sr} \\ 0 & A_{ryt} & A_{rs} & A_{rr} \end{bmatrix} \begin{bmatrix} \Psi_{sy} \\ \Psi_{ry} \\ U_{st} \\ U_{rt} \end{bmatrix} = \begin{bmatrix} W_c i_s \\ i_b \\ 0 \\ 0 \end{bmatrix} \tag{50}$$

Due to core saturation characteristic is a non-linear curve, the matrix Eq. (50) is non-linear and some coefficient matrices depends to the core fluxes density. Thus, ordinary methods cannot be employed for solving Eq. (50). Furthermore, rotor and stator currents are depended to stator three phase source voltages and rotor speed in differential equation forms. Therefore, for detailed analysis of squirrel cage induction machine, it is necessary to solve an electric, mechanic and magnetic algebraic differential equations, simultaneously.

ELECTRICAL VOLTAGE EQUATIONS OF SQUIRREL CAGE INDUCTION MACHINES

Generally, a set of differential equations in an electrical machine is used to describe rates of the electrical and mechanical variables. These equations establish the relationship between fluxes, currents and the three phase source voltage variables. In the next section, rotor and stator voltage relations are be derived.

Stator Voltage Equations

For an induction machine, we can write electrical differential equations in stator windings as:

$$V_s = R_s i_s + \frac{d}{dt}\Lambda_s \tag{51}$$

where V_s and R_s are voltage and stator resistances matrix, respectively which are defined by

$$V_s = \begin{bmatrix} v_a & v_b & v_c \end{bmatrix}^T \tag{52}$$

$$R_s = diag\begin{bmatrix} r_a & r_b & r_c \end{bmatrix} \tag{53}$$

Moreover, Λ_s is linkage flux vector and equals to the product of turn number of stator windings and the phase fluxes. Thus, we can write:

$$\Lambda_s = \begin{bmatrix} \lambda_a & \lambda_b & \lambda_c \end{bmatrix}^T \tag{54}$$

$$\Lambda_s = w_c \Psi_{3\varphi} \tag{55}$$

By substituting Eqs. (34) and (37) into (55), we obtain:

$$\Lambda_s = w_c M_{cf} A_{syt} \Psi_{sy} \tag{56}$$

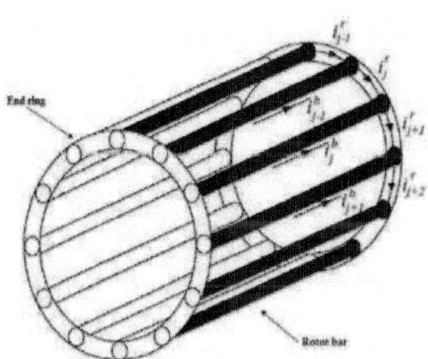

Figure 5: Structure of squirrel cage rotor

Rotor Voltage Equations

Fig. 5 shows a structure of squirrel cage rotor in an induction machine. The rotor topology has l + 1 mesh, thus we can write l + 1 independent differential equations to describe electrical dynamic of rotor variables. Suppose, the rotor rings segments have a symmetrical structure, thus number of l independent equations will be enough for modelling of the rotor dynamics. According to Fig. 1, j^{th} tooth of rotor is surrounded by j^{th} and $(j + 1)^{th}$ rotor bars. Thus, based on currents directions of bars and ring segments which are shown in the Fig. 4; the voltage equation in this loop can be given by:

$$\frac{d}{dt}\phi_j^{rt} = -2i_{j+1}^r r_{j+1}^r - i_j^b r_j^b + i_{j+1}^b r_{j+1}^b \tag{57}$$

This relation can be rearranged in the matrix form and the electrical differential equations for the rotor cage can be obtained as

$$\frac{d}{dt}\Psi_{rt} = -2\mathbf{J}_r \mathbf{R}_r \mathbf{i}_r - \mathbf{A}_{ryt} \mathbf{R}_b \mathbf{i}_b \tag{58}$$

where the vector $\mathbf{i}_r \in \mathbb{R}^{l \times 1}, \mathbf{R}_b \in \mathbb{R}^{l \times l}$ and $\mathbf{R}_r \in \mathbb{R}^{l \times l}$ denote ring segment currents, rotor bar and ring segment resistance matrices, receptively. These vectors are given by:

$$\mathbf{i}_r = \begin{bmatrix} i_1^r & i_2^r & \cdots & i_l^r \end{bmatrix}^T \tag{59}$$

$$\mathbf{R}_r = diag\begin{bmatrix} r_1^r & r_2^r & \cdots & r_l^r \end{bmatrix} \tag{60}$$

$$\mathbf{R}_b = diag\begin{bmatrix} r_1^b & r_2^b & \cdots & r_l^b \end{bmatrix} \tag{61}$$

Also, constant matrix $\mathbf{J}_r \in \mathbb{R}^{l \times l}$ is defined as

$$\mathbf{J}_r = \begin{bmatrix} 0 & 1 & 0 & \cdots & 0 & 0 \\ 0 & 0 & 1 & \cdots & 0 & 0 \\ 0 & 0 & 0 & \cdots & 0 & 0 \\ \vdots & \vdots & \vdots & \ddots & \vdots & \vdots \\ 0 & 0 & 0 & \cdots & 0 & 1 \\ 1 & 0 & 0 & \cdots & 0 & 0 \end{bmatrix} \tag{62}$$

Moreover, there are two current vectors in left hand side of Eq. (58) which have to find a relation between them. For this propose, Kirchhoff's Current Law (KC L) can be used to determined a relation between rotor bars and ring segments currents. By applying KC L to j^{th} node (bar) in the rotor cage (Fig. 4), we can write:

$$i_j^b = i_{j+1}^r - i_j^r \tag{63}$$

Therefore, for l nodes in the rotor cage; Eq. (63) can be expressed in matrix form as:

$$\mathbf{i}_b = \mathbf{A}_{ryt}\mathbf{i}_r \tag{64}$$

By substituting Eqs. (64) into (58), we can get:

$$\frac{d}{dt}\Psi_{rt} = \left(-2J_r\mathbf{R}_r + \mathbf{A}_{ryt}\mathbf{R}_b\mathbf{A}_{ryt}\right)\mathbf{i}_r \tag{65}$$

Up to this stage, two separate sets of non-linear equations are derived to established of algebraical (magnetic) and dynamicalal (electric) models of a squirrel cage induction machine. By combining of Eqs. (50), (51) and (65), the total algebra-differential equations of the system can be augmented as:

$$\begin{bmatrix} w_c M_{cf} A_{syt} p & 0 & 0 & 0 & R_s & 0 \\ 0 & A_{ryt} p & 0 & 0 & 0 & 2J_r R_r - A_{ryt} R_b A_{ryt} \\ A_{\phi sy} + A_{\phi st} A_{syt} & 0 & A_{ust} & 0 & -W_c & 0 \\ 0 & A_{\phi ry} + A_{\phi rt} A_{ryt} & 0 & A_{urt} & 0 & A_{ryt} \\ -A_{syt} & 0 & A_{ss} & A_{sr} & 0 & 0 \\ 0 & -A_{ryt} & A_{rs} & A_{rr} & 0 & 0 \end{bmatrix} \begin{bmatrix} \Psi_{sy} \\ \Psi_{ry} \\ U_{st} \\ U_{rt} \\ i_s \\ i_r \end{bmatrix} = \begin{bmatrix} V_s \\ 0 \\ 0 \\ 0 \\ 0 \\ 0 \end{bmatrix} \tag{66}$$

where p denotes time derivative operator $\left(\frac{d}{dt}\right)$.

Mechanical Differential Equations

Some of the matrix coefficients in Eq.(66) are depended to mechanical instantaneous angle γ. In variable speed conditions, the mechanical variables of the system can be determined by solving differential equations of the rotor angel and speed. Generally, mechanical torque balance equation can be expressed as:

$$J\frac{d\omega}{dt} = T_{em} - T_l \tag{67}$$

Table 1: Parameters of squirrel cage induction machine

Quantity	Symbol	Value
Power	P	1.1kw
Voltage	V	220V
Frequency	f	50Hz
Number of pole	p	2
Stator resistance	r_s	5Ω
Rotor bar resistance	r_b	20μΩ
Rotor ring sector resistance	r_e	1.1μΩ
Inertia moment	j	$0.02 kgm^2$
Number of turns per slot		68
Number of rotor slots		18

in which

$$\frac{d\gamma}{dt} = \omega \tag{68}$$

also, ω, J, T_1 and T_{em} are the rotor angular speed, total inertia on the shaft, load torque and electromagnetic torque, respectively. The electromagnetic torque is depended to mmf along air gap and derivative of air gap permeances with respect to rotor angel (γ) (Ostovic, 1989). For an induction machines, we can express as:

$$T_{em} = \sum_{i=1}^{h}\sum_{j=1}^{l} \left(u_i^{st} - u_j^{rt}\right)^2 \frac{dG_{i,j}(\gamma)}{d\gamma} \tag{69}$$

By substuting Eqs.(7), (10) and (11) into (69), air gap electromagnetic torque can be obtained as:

$$T_{em} = \left(\mathbf{U}_{st}^T - \mathbf{U}_{rt}^T\right) \frac{d}{d\gamma} \mathbf{A}_{sr} \left(\mathbf{U}_{st} - \mathbf{U}_{rt}\right) \tag{70}$$

Therefore, the mechanical non-linear differential equations (67) to (70), with the magnetic non-linear differential-algebraic equations (66), describe the generalized non-linear dynamic model of squirrel cage induction machine. Because of non-linearity of the model, the advantage numerical solution methods must be used. In the next section, some simulation results are presented for demonstration of capability validation of the new model. After that, asymmetrical situations of squirrel cage induction machine such as broken rotor bar fault with saturation effects are analysed and evaluated.

SIMULATIONS AND EXPERIMENT OF RESULTS

Table 1 shows the electrical parameters of the squirrel cage induction machine which is used to obtain numerical simulation and experimental results. Air gap, rotor and stator slots and other main dimensions of the induction machines are presented in Figure (5) and windings connection diagram are assumed similar to Fig. (4). In the next section, some different dynamical conditions are implemented to obtain numerical results of the non-linear model of induction machines.

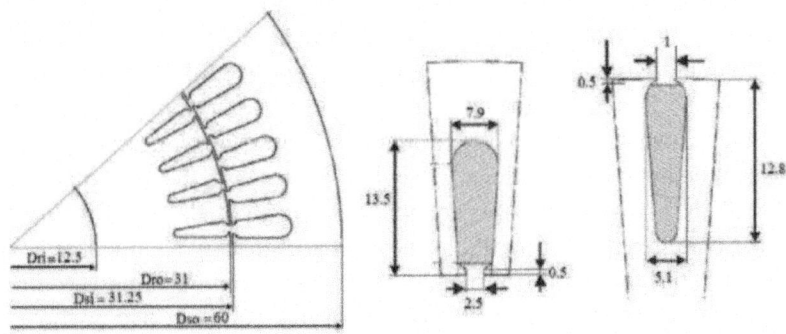

Figure 6: Basic construction and main diameters and stator and rotor slot shapes

Simulation in Healthy Condition

For simulation study, the healthy conditions of a squirrel cage induction machine with saturable iron core are given. At first, free starting-up is considered and at t = 0.3sec, a mechanical load (T_1 = 10Nm) is applied to the rotor shaft. The results of the simulations are shown in Figs. 7 and 8. Speed-torque transient acceleration of the system is presented in Fig. 7-a, that rotor speed reaches to steady state value at about t = 0.2sec. Also, transient electromagnetic torque are shown in Fig. 7-a where low order harmonics are appeared in the electromagnetic torque on the transient duration. This is a common behaviour in the induction machines, which can be obtained from classical method such as two axes theory. But the high order harmonics in transient and steady state are also appeared in the results. At t = 0.3sec load torque changes form zero to T_1 = 10N.m, electromagnetic torque increase by decreasing of the rotor speed. Fig. 7-b shows dynamic of three phase stator currents which decrease by increasing the rotor speed. Other dynamic performance for flux and rotor bar currents are shown in Fig. 8. The results show that, high order harmonics are observed in the rotor and stator teeth fluxes and currents. The rotor teeth and slots are moved opposite the stator teeth and slots, thus radial fluxes will have variable permeances in the air gap. This is caused a slot harmonics appears in the the all variables of the system. Fig 8-a illustrates that, slot harmonics deform the stator teeth flux in the transient and steady state conditions. Because of the machine is symmetric, all variables of machine such as the stator currents and stator teeth flux are balanced.

Simulation in Faulty Condition

In this simulation , analysis of the induction machine with a broken bar is done to determined the steady state performances of the machine. Fig. 9-a shows the steady state current of the stator winding with linear and non-linear core characteristics. The rotor speed and torque are 307.7rad/sec and 10N.m, respectively. Similar to the last conditions, slot harmonics are appeared on the stator currents in the both causes (linear and non-linear magnetic core). However, stator current has low order harmonics with the non-linear iron core. Stator teeth fluxes in two cases are approximately equal. Also, Fig. 9-c presents that the bar currents near to the broken bar are strongly changed. But attention to Fig 9-d, this situation is not appeared in induction machine with non-linear magnetic core. Therefore, disturbances of rotor broken bar in the saturated machines cannot effectively transfer to the stator currents.

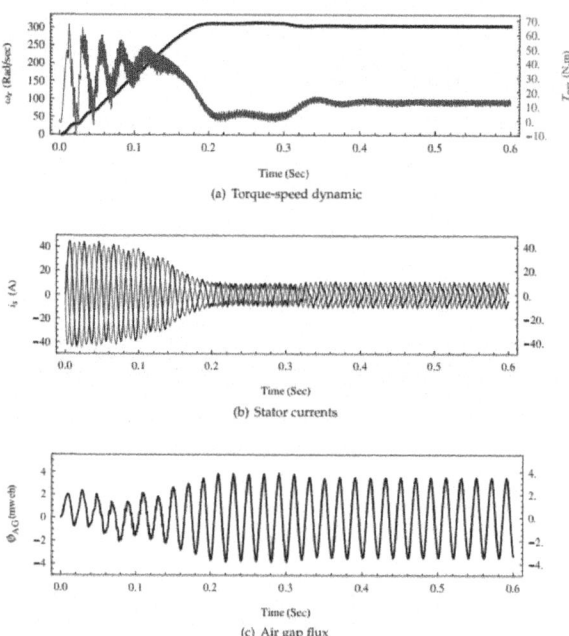

Figure 7: Starting-up transient of induction machine

CONCLUSION

In this chapter, a generalized non-linear dynamical model of the squirrel cage induction machines was presented. In this modelling, magnetic saturation effects in iron core, space harmonic distributed of fluxes in the rotor and stator teeth, the stator windings and rotor bar distribution were considered. Some simulations results showed that, presented model has high accuracy and efficiency for asymmetrical analysis such as broken bar conditions. The effects of broken bar will not clearly appear when the machine core is saturated.

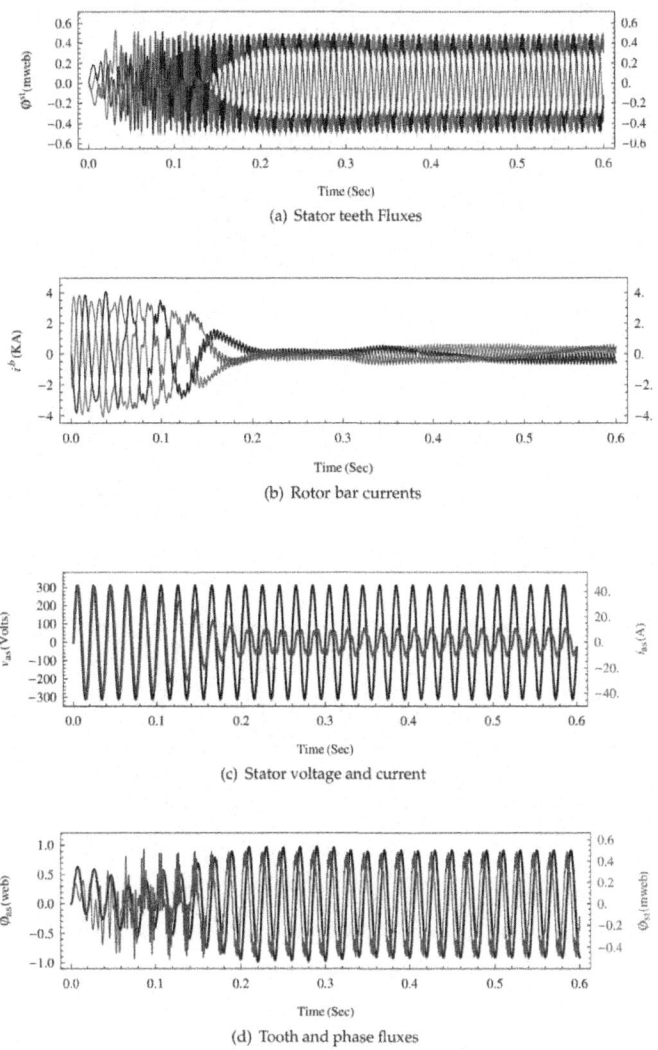

(a) Stator teeth Fluxes

(b) Rotor bar currents

(c) Stator voltage and current

(d) Tooth and phase fluxes

Figure 8: Starting-up transient of induction machine

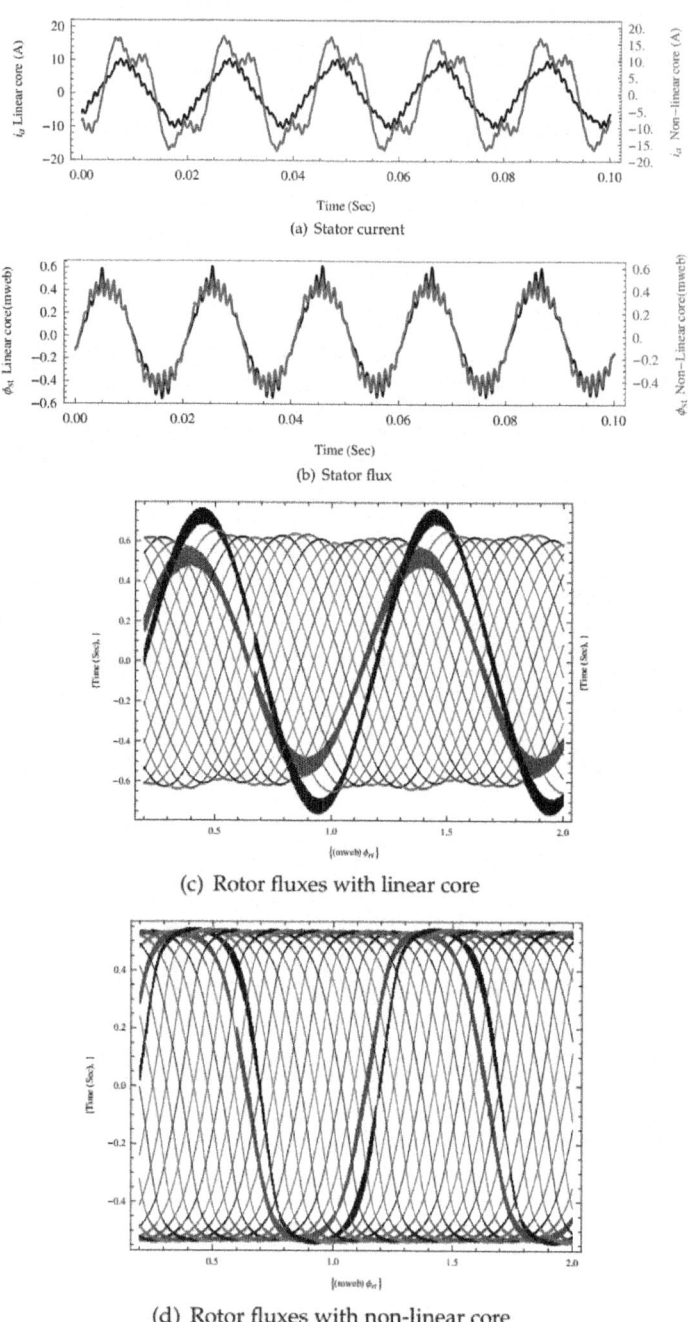

(a) Stator current

(b) Stator flux

(c) Rotor fluxes with linear core

(d) Rotor fluxes with non-linear core

Figure 9: Steady state fluxes and currents of faulty induction machine with a broken bar for linear and non-linear core

REFERENCES

1. Alemi, P. & Nazarzadeh, J. (2006). The Induction machines modeling based on bars current for rotor fault analysis, 21th International Power System Conference, PSC 2006, 477-483, Niroo Research Institute, 13-15 Nov. 2006, Tehran, Iran
2. Al-Shahrani, A. S. (2005). Influence of adjustable speed drive on induction motor fault detection using stator current monitoring. Ph.D. thesis, Oregon State University
3. Bellini, A.; Filippetti, F.; Franceschini, G.; Tassoni, C. & Kliman G. B. (2001). Quantitative evaluation of induction motor broken bars by means of electrical signature analysis.
4. IEEE Transactions on Industry Applications, Vol. 37, No. 5, (Sep./Oct. 2001), 1248-1255 Benbouzid, M. E. H. (2000). A review of induction motors signature analysis as a medium for faults detection.IEEE Transactions on Industry Electronics, Vol. 47, No. 5, (Oct. 2000), 984-993
5. Chen, S. D.; Lin, R. L & Cheng, C. K.(2005). Magnetizing inrush model of transformers based on structure parameters.IEEE Transactions on Industry Electronics, Vol. 20, No. 3, (July 2005), 1947-1954
6. Faiz, J.; Sharifian, M. B. B.; Feyzi, M. R.; & Shaarbafi, K. (2002). A complete lumped equivalent circuit of three-phase squirrel-cage induction motors using two-dimensional finite-elements technique. IEEE Transactions on Energy Conversion, Vol. 17, No. 3, (Spet. 2002), 363-367
7. Gautschi W. (1964). Error function and Fresnel integrals, Handbook of Mathematical Functions, NBS Appl. Math. Series, Vol. 55, U.S. Government Printing Office, Washington, D.C.
8. Jeong, J. H.; Lee E. W. & Cho, H. K. (2003). Analysis of transient state of the squirrel cage induction motor by using magnetic equivalent circuit method, Sixth International Conference on Electrical Machines and Systems, Vol. 2, 720-723, 2003
9. Jung, J. H.; Lee, J. J. & Kwon, B. H. (2006). Online diagnosis of induction motors using MCSA.
10. IEEE Transactions on Industry Electronics, Vol. 53, No. 6, (Dec 2006), 1842-1852 Krause, P. C.; Wasynczuk, O. & Sudhoff, S. D. (2002). Analysis of Electric Machinery and Drive
11. Systems (2nd Edition), Wiley-IEEE Press, ISBN 047114326X, New York.

12. Luos, X.; Liao, Y.; Toliyaty, H.; El-Antably, A. & Lipos, T. A. (1995). Multiple coupled circuit modeling of induction machines. IEEE Transactions on Industry Applications, Vol. 31, No.2, (Mar./Apr. 1995), 311-318
13. Mirafzal, B.& Demerdash, N. A. O. (2004). Induction machine broken-bar fault diagnosis using the rotor magnetic field space-vector orientation. IEEE Transactions on Industry Applications, Vol. 40, No. 2, (Mar./Apr. 2004), 534-542
14. Mirafzal, B.& Demerdash, N. A. O. (2008). Induction machine broken bar and stator short-circuit fault diagnostics based on three-phase stator current envelopes. IEEE Transactions on Industry Applications, Vol. 55, No. 3, (March 2008), 1310-1318
15. Muñoz, A. R. & Lipo, T. A. (1999). Complex vector model of the squirrel-cage induction machine including instantaneous rotor bar currents. IEEE Transactions on Industry Applications, Vol. 35, No. 6, (Nov./Dec. 1999), 1332-1340
16. Ostovic, V. (1989). A novel method for evaluation of transient states in saturated electric machine. IEEE Transactions on Industry Applications, Vol. 25, No. 1, (Feb. 1989), 96-1000 Ostovic, V. (1989). Dynamics of Saturated Machines, Springer-Verlag, ISBN 0387970797, New
17. York
18. Siddique, A.; Yadava, G. S. & Singh, B. (2005). A review of stator fault monitoring techniques of induction motors. IEEE Transactions on Energy Conversion, Vol. 20, No. 1, (March 2005), 106-114
19. Sprooten, J. (2007). Finite element and electrical circuit modeling of faulty induction machines study of internal effects and fault detection techniques. Ph.D. thesis, Department of Bio, Electro and Mechanical Systems (BEAMS), University Libre de Bruxelles
20. Su, H. & Chong, K. T. (2007). Induction machine condition monitoring using neural network modelling. IEEE Transactions on Industry Applications, Vol. 54, No. 1, (Feb. 2007), 241-249

Chapter 6

MINIMIZATION OF LOSSES IN CONVERTER-FED INDUCTION MOTORS – OPTIMAL FLUX SOLUTION

Waldiberto de Lima Pires, Hugo Gustavo Gomez Mello, Sebastião Lauro Nau, and Alexandre Postól Sobrinho

WEG Equipamentos Eletricos S.A. – Motores Research and Development of Product Department Av. Pref. Waldemar Grubba, 3000 – malote 41 Jaraguá do Sul, SC - 89256-900 Brazil

INTRODUCTION

When a TEFC induction motor fed by static frequency converter drives a load which demands constant torque throughout the operation range, the low speeds are thermally critical, because the motor losses (heat sources) do not vary much as a function of the speed, but the ventilation efficiency decreases as the operation speed falls down, since the fan is installed on the very motor shaft. In such cases, when operating at low speeds, the temperature rise often exceeds the limits of the motor thermal class due to the lack of cooling. In order to prevent this problem the industry has traditionally adopted one of the following solutions: independent ventilation (a small auxiliary motor is used to exclusively drive the fan that provides the main motor cooling) or oversizing (the motor used in the application provides a higher torque than the rated load demand). However, neither one nor the other of these two options are attractive, as both, besides increasing the space required for the installation, still increase the motor price [1].

Frequency converters usually apply to the motor a constant voltage/frequency ratio throughout the operation range, so that no loss control is provided. But the study on the motor losses composition and its relation with voltage, frequency, magnetic flux and current, allied with the study on the influence of the ventilation on the temperature rise of the motor, has led to an optimal voltage/frequency ratio, which minimizes the total motor losses at each speed [2]. This way, by implementing the automatic control of the voltage/frequency ratio in the converter, the motor loss minimization can be automatically obtained throughout the frequency range, so that the motor temperature rise is kept within the thermal class limits even at low speeds with reduced ventilation.

DETERMINATION OF LOSSES

The fast growth of the number of industrial applications using static frequency converters recently observed in variable speed drives has encouraged the meticulous study of losses in magnetic materials under PWM supply by several researchers [3, 4, 5, 6, 7, 8]. They have shown that such losses depend on a number of control parameters, such as the modulation index, the number of levels and pulses of the frequency converter and the duration of PWM signal pulses. On the other hand, Boglietti et alli have concluded that the flux waveforms resulting from PWM supply differ from those resulting from sinusoidal supply just except for a small ripple, which depends on the switching frequency, and that above approximately 5 kHz the iron losses can be considered independent of this parameter [9]. Such studies represent the first step towards the understanding of the losses behavior in electric motors under PWM supply, that involves a higher degree of complexity and is not restricted exclusively to the magnetic materials issue, but includes also additional losses in the conductors and due to the cooling system and depends, besides the control parameters, on some machine design parameters, such as the flux density, the lamination geometry and the connection of windings [3, 10, 11, 12], as well as on other variables inherent to the manufacturing process [13].

For the purposes of this study, however, the analysis of the motor losses can be simplified, so that it is enough to separate the total motor losses in three key components:

$$P = P_{fe} + P_j + P_{mec} \qquad (1)$$

where:

P_{fe} – Iron losses, which depend on the flux density (or magnetic induction), the frequency and the quality of the magnetic material.

P_j – Losses by Joule effect, which depend on the currents flowing through the stator windings and the rotor bars.

P_{mec} – Mechanical losses due to cooling system (fan coupled to the shaft) and friction, which depend on the speed.

The iron losses are classically considered as being composed of two portions: Hysteresis losses (pH) and induced eddy current (Foucault) losses (pF). For a lamination sample tested in Epstein Frame with sinusoidal supply, the hysteresis losses are directly proportional to the frequency (f) and to the square of the magnetic induction (B^2), while the eddy current losses are proportional to the square of both the frequency (f^2) and the magnetic induction (B^2), especially for induction values above one Tesla (1 T). However, in the induction motor the iron losses present a much more induction-dependent behavior than the quadratic ratio obtained with normalized samples of the magnetic material in Epstein Frame tests. In low-voltage three-phase induction motors manufactured with fully processed steel laminations tested under different saturation levels, iron losses presented a dependance on the induction close to B^4 for inductions above 1,2 T (usual value for industrial motors), as presented in Fig. 1.

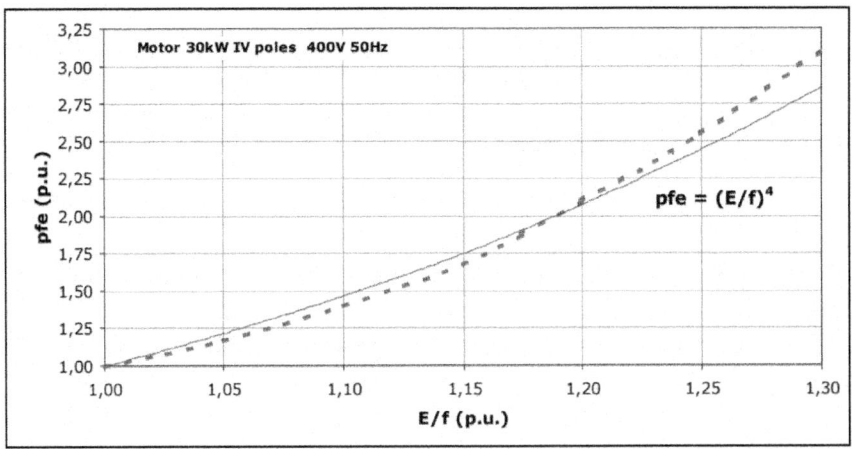

Figure 1: Iron losses x Magnetic induction for an industrial three-phase induction motor

The fundamental theory of the electric machines shows that the torque provided by the induction motor is directly proportional to the product between the magnetic flux and the electric current [14, 15]. Then in order to keep a constant torque, if the flux increases the current can decrease (and vice-versa). As the Joule losses are directly proportional to the square of the current, these losses can be considered as inversely proportional to the square of the magnetic flux. From the Faraday-Lenz law of induction, one can easily demonstrate that the magnetic flux in the motor is directly proportional to the ratio between the electromotive force (E) and the frequency. Considering the steady-state model of equivalent circuit of the induction motor per-phase (Fig. 2), it can be noted that at the base frequency the voltage drop in the primary impedance has little significance, so that the flux can be considered as proportional to the *V1/f* (voltage/frequency) ratio.

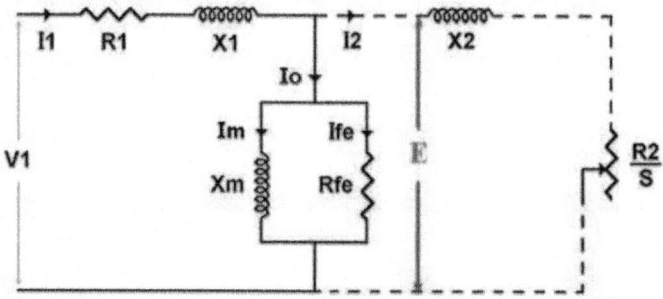

Figure 2: Steady state equivalent circuit of the induction motor per-phase

For low operating frequencies, however, in which the input voltage is reduced, the voltage drop in the primary resistance becomes important and can be no longer despised. By neglecting the influence of the primary reactance, electromotive force E is given by

$$E = V - R_1 I_1 = V - \Delta V \quad (2)$$

The voltage drop at the stator branch (ΔV) then depends directly on the stator current ($I1$). As Fig. 2 shows, the motor current can be decomposed into two components: one concerning magnetization and the other concerning torque production.

$$I_1 = \sqrt{I_0^2 + I_2^2} \quad (3)$$

Taking rated voltage as the base, E/f ratio per unit can be written as:

$$\frac{E}{f} = \frac{V}{f} - \frac{\Delta V_n}{f} \sqrt{\frac{k_T^2 . k_m^2}{\left(\frac{E}{f}\right)^2} + k_{i0n}^2 . k_{i0}^2} \quad (4)$$

where:

ΔVn - Voltage drop per unit with rated frequency and load.

f – Motor operating frequency per unit, considering rated frequency fn as the base. In (4), the square root results in a correction factor, which is function of the motor current and whose terms are explained in the following paragraphs.

It should be taken into account that, as the frequency (and consequently, the rotation) is reduced, the mechanical losses decrease in a nearly cubic proportion to it (f^3). The mechanical losses do not affect the iron losses, but they act as an additional load to the motor, therefore they must be considered as a torque to be added to the rated torque available on the shaft. Its reduction implies current reduction and so reduction of Joule losses in the conductors (Pj).

Thus it is possible to rewrite the induction motor total losses per unit p, for operation with variable voltage and frequency, as follows:

$$p = p_{in}\left[\frac{k_T^2 \cdot k_m^2}{\left(\frac{E}{f}\right)^2} + k_{i0n}^2 \cdot k_{i0}^2\right] + p_{Hn}\left(\frac{E}{f}\right)^4 f + p_{Fn}\left(\frac{E}{f}\right)^4 f^2 \quad (5)$$

where:

pin – Total Joule losses with the motor operating at rated conditions of load, voltage and frequency.

pHn – Total hysteresis losses with the motor operating at rated conditions of load, voltage and frequency.

pFn – Total eddy current losses with the motor operating at rated conditions of load, voltage and frequency.

Motors manufactured with low loss magnetic core (fully processed silicon steel) operating at rated conditions typically present values of 80%, 12% and 8%, for parameters *pin*, *pHn* and *pFn*, respectively. The remaining parameters of (5) will be opportunely explained ahead. The term of (5) in brackets refers to the motor Joule losses and depends on the total motor current. The second and third terms refer to the motor iron losses for hysteresis and eddy currents, respectively. The magnetic induction was conveniently replaced by the E/f ratio. There is no explicit term for the motor total mechanical losses in the equation, because they are embedded in the first term, in accordance with what was mentioned before, by means of the factor km defined below:

$$k_m = \left(\frac{1 + p_{mn} f^3}{1 + p_{mn}}\right) \quad (6)$$

where *pmn* is the mechanical losses at rated speed referred to the rated output power *Pn*. The aim of this study is to minimize the motor losses, in order to reduce its temperature rise, so that the need of both the torque reduction (oversizing) and the use of independent ventilation can be prevented. In (5), this is considered by means of the derating factor kT, which will be addressed later on this paper. It should be noted that the torque affects only the current-dependent losses, not influencing the iron losses. Therefore, km and kT are torque correction factors required to compensate for the effects of the speed variation, which influences the portion of losses related to the load current.

$Ki0n$ is the no-load current factor, defined by (7).

$$k_{i0n} = \frac{I_o}{I_n} \qquad (7)$$

where:

Io – No-load current under rated voltage and frequency. In – Full-load current under rated voltage and frequency.

Due to the non-linearity of the magnetization curve of the laminations, the E/f ratio increase causes the no-load current to increase according to (8). This peculiar behavior of the no-load current was observed experimentally (Fig. 3), and is taken into account in (5) by means of the factor $ki0$.

$$k_{i0} = \left(\frac{E}{f}\right)^{3,4} \text{ para } \frac{E}{f} \geq 1$$

$$k_{i0} = \left(\frac{E}{f}\right) \text{ para } \frac{E}{f} < 1 \qquad (8)$$

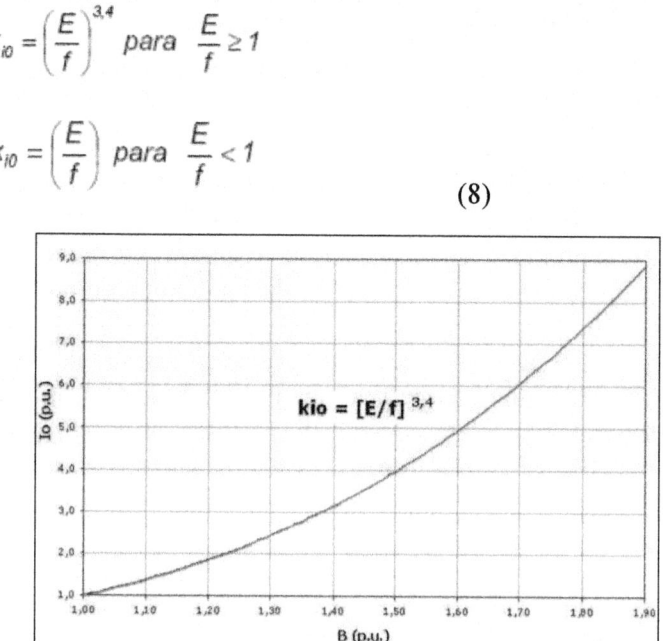

Figure 3: Magnetizing current x E/f

MINIMIZATION OF LOSSES

The analysis of (5) shows that the induction motor global losses depend on both the operation frequency and the induction (or magnetic flux). Then the values of V/f that minimize the motor global losses change with the operation frequency, so that it is necessary to find the minimum losses at each frequency, with different values of V/f. Fig. 4 shows the total losses calculated as function of the frequency for various values of the V/f ratio.

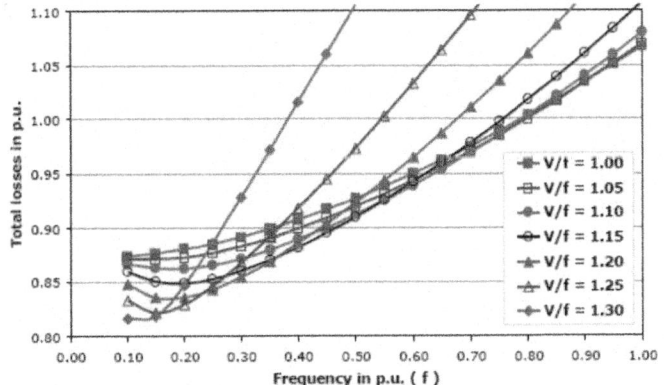

Figure 4: Total losses x frequency curve for several V/f ratios at rated torque

Fig. 5 derives from the family of curves above and represents the V/f ratios theoretically obtained, which minimize the total losses of the motor at each operation frequency.

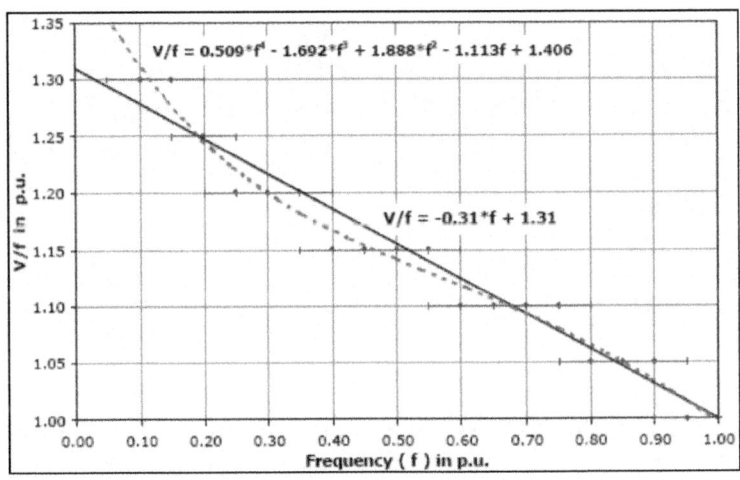

Figure 5: V/f x frequency curve for minimization of total losses

INFLUENCE OF VENTILATION REDUCTION

Thermal calculations implemented for a number of motors of distinct frame sizes and power ratings considering speed variation, combined with experiments and tests performed with several motors at rated load, varying separately the fan speed from zero to base speed, led to the conclusion that TEFC three-phase motors of a wide output range present a similar thermal behavior. Fig. 6 represents the temperature rise per unit of low-voltage 4-pole cage induction motors manufactured with die cast iron frame as a function of the fan speed per unit.

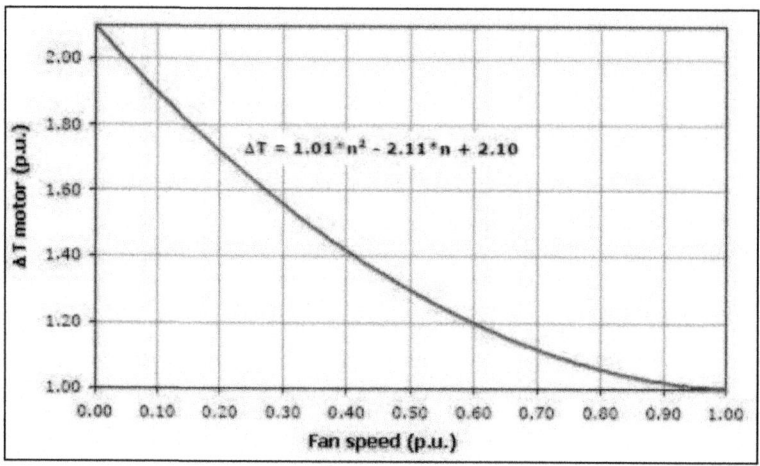

Figure 6: Temperature rise x fan speed at rated load

In this approach, for each desired value of frequency and for each value of V/f ratio according to Fig. 5, the total losses p are calculated according to (5). Fig. 6 determines the influence of the ventilation reduction on the motor temperature rise ($Tmotor$). So in order for the required motor temperature rise to be assured, it is necessary to calculate a new value of p, henceforth referred to as p', according to (9).

$$p' = \left(\frac{\Delta T}{\Delta T_{motor}} \right) p_n \qquad (9)$$

where:

p' – total motor losses for the required temperature rise, considering the ventilation reduction.

pn - total motor losses at rated conditions.

If, for instance, the maximum required temperature rise is the limit of the insulation system thermal class, then $\Delta T = \Delta Tclass$. If, otherwise, the maximum required temperature rise is

ΔTn, then $\Delta T=1$.

Once known the total losses p' that will cause the required temperature rise in the motor with reduced ventilation, then it is possible to calculate the convenient derating factor using (10):

$$k_T = \frac{k_{HVF}}{k_m}\left(\frac{E}{f}\right)\sqrt{\frac{p'-p_{Hn}\left(\frac{E}{f}\right)^4 f - p_{Fn}\left(\frac{E}{f}\right)^4 f^2}{p_{in}} - k_{i0n}^2 \cdot k_{i0}^2} \qquad (10)$$

where $kHVF$ is the harmonic voltage factor as defined by NEMA [16]. It was placed in the equation originally conceived as (5) for the influence of the PWM supply voltage harmonics to be also considered on the motor temperature rise. For most of the modern static frequency converters $kHVF$ is 0.95.

As a consequence of the cooling reduction, kT is usually lower than 1. However, if the minimized total losses are such that reduce the motor temperature rise even with poor ventilation, kT can be higher than 1. Similarly, if kT is calculated for an insulation class temperature rise, it will be normally higher than 1 if the rated motor temperature rise is much below the insulation class temperature rise limit.

Fig. 7 presents an example of loss reduction achieved with the proposed method. A three-phase, 30 kW, 4-pole induction motor was tested at constant rated torque within the frequency range from 0.1 to 1.0 (p.u.). The results are presented for three different situations: calculation with constant flux, calculation with optimal flux and testing with optimal flux.

The loss reduction obtained with the proposed technique, as shown in Fig. 7, results in a better thermal performance of the motors operating with optimal flux. Comparing the motor temperature rises when operating at constant flux condition to those when operating at optimal flux condition, it is remarkable a behavior similar to that outlined in Fig. 8, as can be checked in the experimental results presented in section VI.

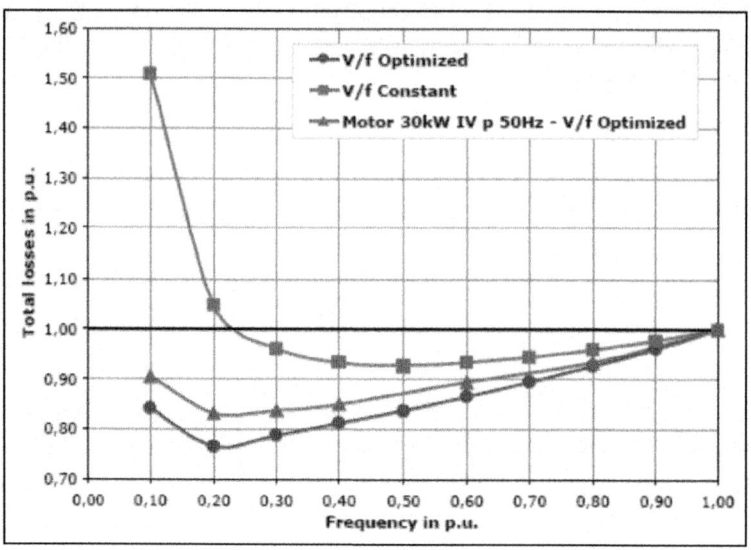

Figure 7: Total losses (p.u.) x frequency (p.u.)

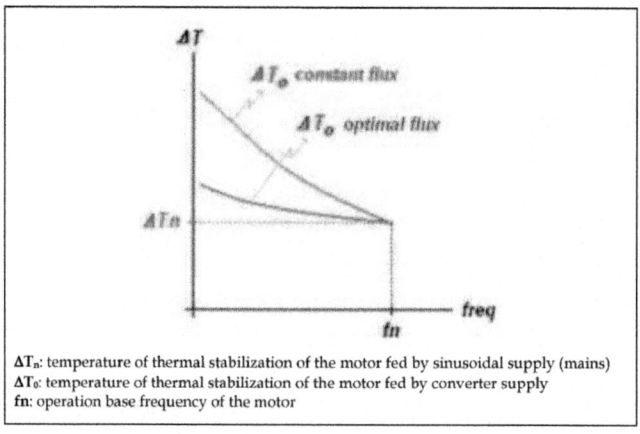

Figure 8: Temperature rise x operation frequency, sketch of the motor performance under different flux conditions

VALIDATION AND IMPLEMENTATION OF THE OPTIMAL FLUX CURVE

To validate the proposed technique, temperature rise tests with speed/ventilation variation were accomplished for a wide range of industrial motor ratings. This way, it was possible to compare the thermal performances of converter-fed motors when under constant flux (rated losses) and optimal flux (minimum losses) conditions.

Before the implementation of the automatic function for optimal flux selection by the converter, drives with suitably modified softwares were used, so that specific flux values could be manually adjusted. This allowed the practical correction of the curve obtained by means of mathematical calculations and the finding of the actual optimal flux curve that was implemented in the converters used in the tests, into which were incorporated the automatic function for the optimal flux setting.

As the sensorless vector control enables the magnetic flux of the motor to be directly altered, this was the control type employed in all tests with converter. The switching frequency used in all tests with converter was 2.5 kHz.

A. NEMA High Efficiency (NHE) Motors

The motors to be tested were selected considering the worst horsepower/frame size ratios criteria. The following machines were used in the tests, all of them 4-pole (predominant polarity in low-voltage industrial applications) and all of them with class F insulation: 5 hp (NEMA 184T); 20 hp (NEMA 256T); 50 hp (NEMA 326T) and 150 hp (NEMA 444T). Occasionally, for investigation of specific issues, tests were also realized with motors that are not related above.

It should be noted that the loss minimization technique was conceived and developed specifically for low operation frequencies (below 0.5 p.u.). Thus, at base (rated) frequency, in this case 60 Hz, the flux value for the minimum losses condition is the rated flux. The tests were conducted with full load being continuously applied until the motor thermal stabilization.

B. NEMA Premium Efficiency (NPE) Motors

Considering that NEMA Premium Efficiency motors tend to present lower temperature rises than NEMA High Efficiency motors, few NPE motors were tested just for the effectiveness of the proposed loss minimization technique to be corroborated with such motors. Despite that tests were conducted with 2- and 4-pole motors of very distinct horsepower rates, in order to extend the solution also for machines of other polarities. The following motors were tested: 5 hp (NEMA 184T) and 150 hp (NEMA 444T).

The tests were conducted in a way similar to the followed for NHE motors: continuous application of full load until the thermal stabilization of the motor.

C. EFF1 (IE3) Motors

The several motors tested were chosen according to the criterion of the most critical frame size/horsepower ratios. They were all 4-pole machines, since this polarity is typical in low-voltage applications involving speed variation. Motors with base frequencies of both 50 and 60 Hz were tested, for a wider verification of the proposed solution, as listed below:

- f_{base} = 60 Hz: 3 hp (IEC 90L), 12.5 hp (IEC 132M), 50 hp (IEC 200L), 75 hp (IEC 225SM) and 150 hp (IEC 280 SM).
- f_{base} = 50 Hz: 2hp (IEC 90L), 10 hp (IEC 132M), 40 hp (IEC 200L), 75 hp (IEC 225SM) and 150 hp (IEC 280 SM).

Regarding the temperature rise tests, it should be noted that at 50 or 60 Hz the motors were always fed by the mains (AC power line), that is, sinusoidal supply was used, while all the other tests were accomplished with the motors fed by converter in optimal flux condition for loss minimization. Due to the fact that EFF1 (IE3) motors usually present lower efficiency levels than NHE and NPE motors, they tend to operate a little bit warmer than the latter. Because of that, unlike the tests performed with NHE and NPE motors, sometimes derating factors were applied in the tests with EFF1 (IE3) motors at low frequencies. However, in all cases the tests were performed following S1 duty as well.

EXPERIMENTAL RESULTS

The results contained in Tables I and II evidence the effectiveness of the proposed loss minimization technique for NEMA High Efficiency and NEMA Premium Efficiency motors: all the motors tested with optimal flux got the thermal stabilization at a lower temperature than with constant flux, at all frequencies analysed. Furthermore, Tables I and II show that, when operating at minimum loss condition, NHE and NPE motors can provide the rated torque continually throughout the operation range, even at low frequencies.

Tables I and II present some lacking results of low frequency (typically 5 Hz or below) tests. In these cases, the respective temperature rise test at full load could not be concluded, either because the motor was already running too hot, forcing the interruption of the test before thermal stabilization was reached, or because the dynamometer could no longer apply the required test load after a period, due to its functioning principle (braking/loading caused by the Foucault effect) and given the increased slip of the motor when it is heated. Temperature rise tests that could not be concluded properly with constant flux, but could be normally performed in the optimal flux condition, are themselves evidences of the thermal performance improvement of the motor, provided by the loss minimization technique.

Table I: Temperature rise tests results (K) 4-pole nhe motors – s1 duty

Source	Operation Frequency	Flux	5 hp	20 hp	50 hp	150 hp
Mains supply	60 Hz sinusoidal	constant	50.3	48.1	53.8	67.7
Converter	60 Hz	constant	51.3	59.1	77.7	86.6
		optimal				
	15 Hz	constant	72.8	79.8	122.3	100.3
		optimal	72.8	62.4	79.8	82.6
	10 Hz	constant	101.0	104.7		125.3
		optimal	77.8	67.7	88.1	83.2
	5 Hz	constant				
		optimal	82.8	80.6	115.8	97.4

Table II: Temperature rise tests results (K) npe motors – s1 duty

Source	Operation Frequency	Flux	5 hp – 4p	150 hp – 4p	5 hp – 2p
Mains supply	60 Hz sinusoidal	constant	28.4	66.5	31.1
Converter	60 Hz	constant	32.5	*	*
		optimal		*	*
	30 Hz	constant	39.9	*	*
		optimal	36.5	*	*
	10 Hz	constant	65.5	*	40.2
		optimal	55.0	*	29.5
	5 Hz	constant		*	
		optimal	67.9	*	82.1
	4 Hz	constant		*	*
		optimal	100.0	*	*
	3 Hz	constant			*
		optimal	*	87.4	*

*Tests that have not been performed

The analysis of the data presented in the tables above by means of graphs makes the interpretation of results easier. Fig. 9 evidences the advantages of using the optimal flux technique with a NHE, 20 hp, 4-pole motor.

Some results of temperature rise tests realized with EFF1 (IE3) motors under optimal flux condition are graphically presented in Figs. 10 to 13.

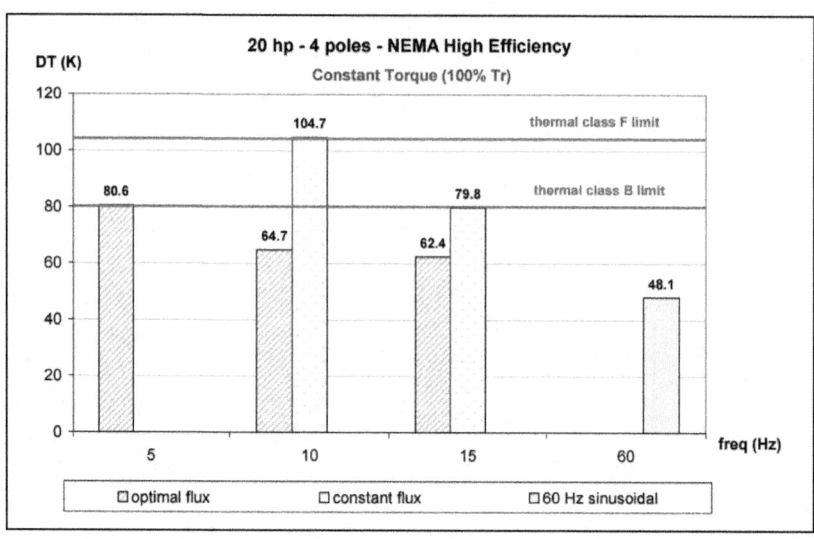

Figure 9: Examples of results of temperature rise tests accomplished with NHE motor

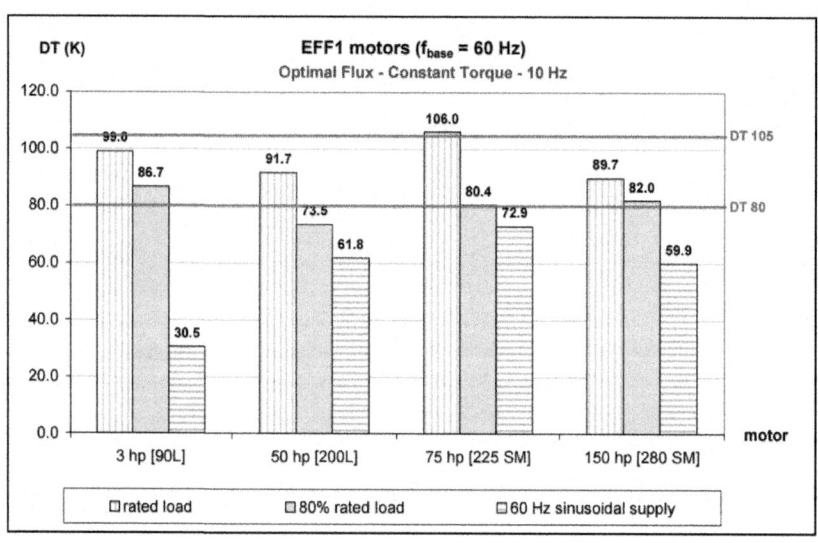

Figure 10: Results of temperature rise tests conducted with EFF1 (f_{base} = 60 Hz) motors running at 10 Hz

Minimization of Losses in Converterfed Induction Motors – Optimal... 125

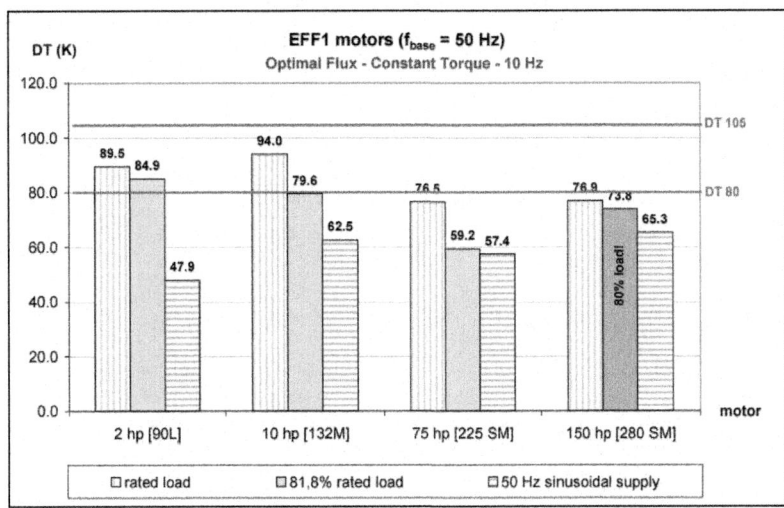

Figure 11: Results of temperature rise tests conducted with EFF1 (f_{base} = 50 Hz) motors running at 10 Hz

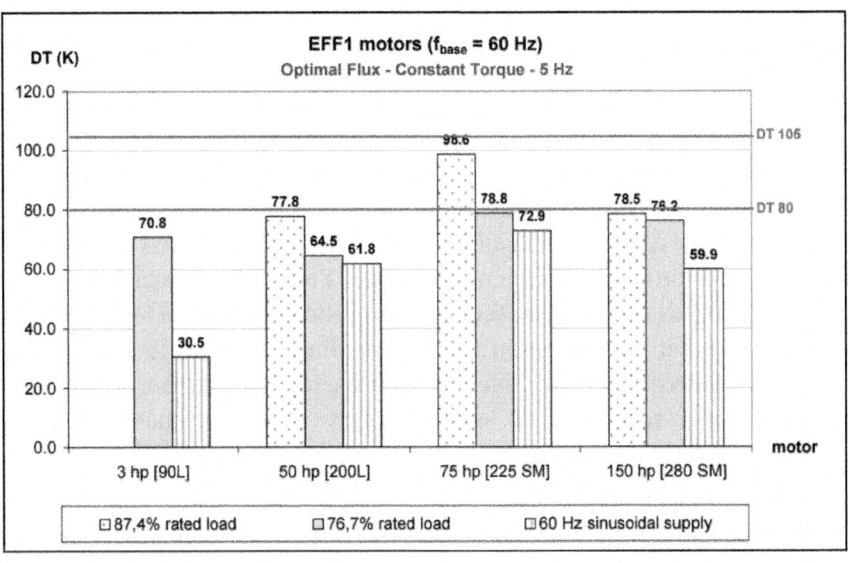

Figure 12: Results of temperature rise tests conducted with EFF1 (f_{base} = 60 Hz) motors running at 5 Hz

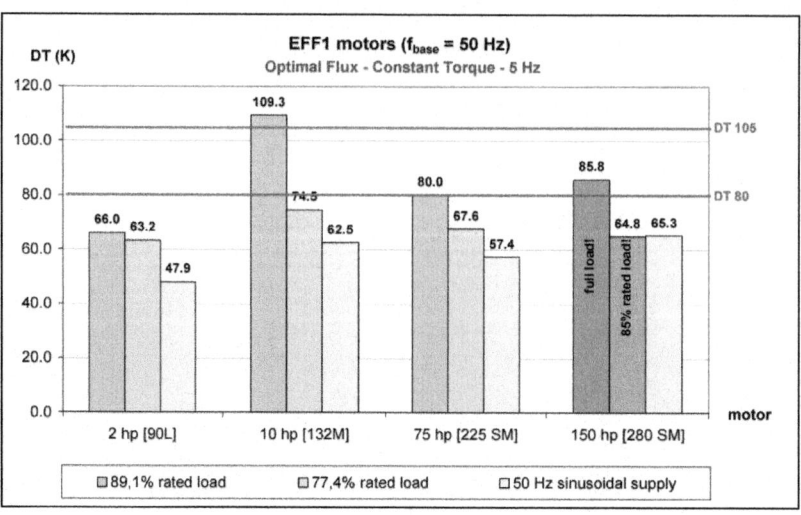

Figure 13: Results of temperature rise tests conducted with EFF1 (f_{base} = 50 Hz) motors running at 5 Hz

CONCLUSIONS

The study on the behavior of 3-phase induction low voltage motors' losses as a function of parameters such as load, operation frequency and magnetic induction (flux density), brought about the conception of an equation for magnetic flux that minimizes the total motor losses in each operation frequency. This is very interesting, given that variable speed drives usually involve operation throughout a speed range. Frequency reduction causes the motor iron losses to decrease, thus allowing the magnetic induction (or flux) to be increased at low operation speeds, enabling the motor to provide torque with lower current. The flux increase results from an increment in voltage/frequency ratio, accomplished in a growingly more accentuated manner the lower the operation frequency.

The mathematical equation that has been created was implemented in commercial static frequency converters and the study validation was carried out by means of the analysis of specific motor lines and the execution of tests with a number of frequency converters and induction motors of distinct sizes, using the optimal flux ratios theoretically determined. The tests led to the definition of the actual optimal flux curve as well as new derating curves valid for motors working under optimal flux condition, which evidence the advantages provided by the proposed technique (Fig. 14). According to Fig. 14, it is possible to note that if temperature rise of thermal class F (105°C) is permitted on the motor windings, the loss minimization technique prevents the need of torque reduction for operation of the motor at low frequencies (until 0.1 p.u.). If temperature rise of thermal class B (80°C) is required, the loss minimization technique allows for the application of a smoother torque derating than the needed for the motor running with constant flux and rated losses.

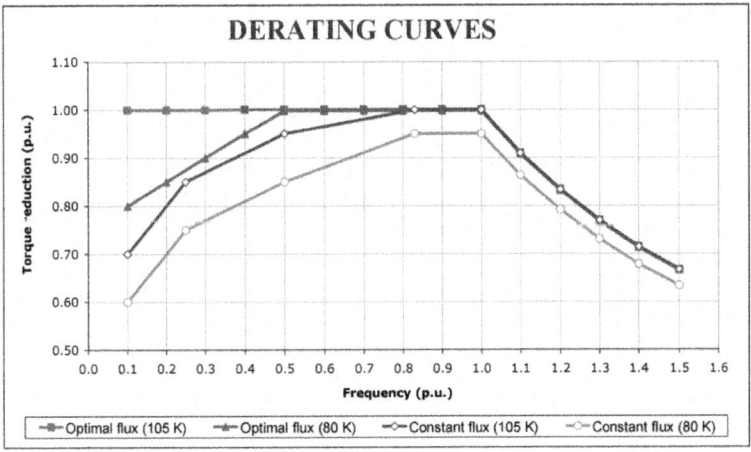

Figure 14: The new derating (torque reduction) curves proposed evidence the advantages provided by the optimal flux solution

The implementation of the optimal flux curve into commercial frequency converters resulted in the development of a solution (converter + motor) to optimize/improve the variable speed drives with constant torque loads.

The next step of this development (ongoing) is improving the Optimal Flux Solution, so that it can be advantageously used with any type of load driven by converter-fed induction motors.

REFERENCES

1. PCT/BR2004/000074, "Static frequency converter with automatic function for optimizing magnetic flux and minimizing losses in electric induction motors".
2. S. L. Nau and A. P. Sobrinho, "Optimal voltage/frequency curve for inverter-fed motor". Proceedings of the 3rd International Conference on Energy Efficiency in Motor Driven Systems (EEMODS), Treviso, Italy, 2002.
3. R. Kaczmarek; M. Amar and F. Protat, "Iron loss under PWM voltage supply on Epstein frame and in induction motor core". IEEE Transactions

on Magnetics, Vol. 32, No.1, January 1996.
4. Boglietti; P. Ferraris; M. Lazzari and F. Profumo, "Iron losses in magnetic materials with six-step and PWM inverter supply". IEEE Transactions on Magnetics, Vol. 27, No. 6, November 1991.
5. L. T. Mthombeni and P. Pillay, "Core losses in motor laminations exposed to high-frequency or nonsinusoidal excitation". IEEE Transactions on Industry Applications, Vol. 40, No. 5, September-October 2004.
6. M. S. Lancarotte; C. Goldemberg and A. A. Penteado Jr, "Estimation of FeSi core losses under PWM or DC bias ripple voltage excitations". IEEE Transactions on Energy Conversion, Vol. 20, No. 2, June 2005.
7. J. Moses and N. Tutkun, "Investigation of power loss in wound toroidal cores under PWM excitation". IEEE Transactions on Magnetics, Vol. 33, No. 5, September 1997.
8. Boglietti; M. Chiampi; M. Repetto; O. Bottauscio and D. Chiarabaglio, "Loss separation analysis in ferromagnetic sheets under PWM supply". IEEE Transactions on Magnetics, Vol. 34, No. 4, July 1998.
9. Boglietti; P. Ferraris; M. Lazzari and M. Pastorelli, "About the possibility of defining a standard method for iron loss measurement in soft magnetic materials with inverter supply". IEEE Transactions on Industry Applications, Vo. 33, No. 5, September-October 1997.
10. Cester; A. Kedous-Lebouc and B. Cornut, "Iron loss under practical working conditions of a PWM powered induction motor". IEEE Transactions on Magnetics, Vol. 33, No. 5, September 1997.
11. A.Ruderman and R. Welch, "Electrical machine PWM loss evaluation basics". Proceedings of the 4th International Conference on Energy Efficiency in Motor Driven Systems (EEMODS), Heidelberg, Germany, 2005.
12. M. Sokola; V. Vuckovic and E. Levi, "Measurement of iron losses in PWM inverter fed induction machines". Proceedings of the 30th Universities Power Engineering Conference (UPEC), London, UK, 1995.
13. A.C. Smith and K. Edey, "Influence of manufacturing processes on iron losses". Proceedings of the 7th International Conference on Electrical Machines and Drives - Conference Publication No. 412. September 1995.
14. K. Kostenko e L. Piotrovski, Máquinas Eléctricas – Volume II – Máquinas de Corrente Alternada (tradução de original russo), Ed. Lopes da Silva, Porto, Portugal, 1979.
15. P. C. Krause, Analysis of Electric Machinery, McGraw Hill, New York, USA, 1996.

16. NEMA Standard MG1-2003, Part 30 – Application considerations for constant speed motors used on a sinusoidal bus with harmonic content and general purpose motors used with adjustable-voltage or adjustable-frequency controls or both.

Chapter 7

SENSORLESS VECTOR CONTROL OF INDUCTION MOTOR DRIVE - A MODEL BASED APPROACH

Jogendra Singh Thongam[1] and Rachid Beguenane[2]

[1]Department of Renewable Energy Systems, STAS Inc., Chicoutimi, QC
[2]Department of ECE, Royal Military College, Kingston, ON Canada

INTRODUCTION

Induction machines are more rugged, compact, cheap and reliable in comparison to other machines used in similar applications. Vector controlled induction motor drive outperforms the dc motor drive because of higher transient current capability, increased speed range and lower rotor inertia.

Sensors widely used in electric drives degrade the reliability of the system especially in hostile environments and require special attention to electrical noise. Moreover, it is difficult to mount sensors in certain applications in addition to extra expenses involved. Therefore, a lot of researches are underway to develop accurate speed estimation techniques. With sensorless vector control we have a decoupled control structure similar to that of a separately excited dc motor retaining the inherent ruggedness of the induction motor at the same time. Speed sensorless control technique first appeared in (Abbondante & Brennen, 1975). The commonly used methods for speed estimation are Model Reference Adaptive System (MRAS) (Schauder, 1992; Tajima & Hori, 1993; Peng & Fukao, 1994; Choy et al., 1996), Neural Networks (Simoes & Bose, 1995; Fodor et al., 1995; Ben-Brahim & Kudor, 1995; Kim et al., 2001; Toqeer & Bayindir, 2003; Haghgoeian et al., 2005), Extended Kalman Filter (EKF) (Kim et al.,1994, Comnac et al., 2001; Ma & Gui, 2002; Du et al., 1995; Thongam & Thoudam, 2004) and Nonlinear Observer (Bodson et al., 1995; Liu et al., 2001, Pappano et al., 1998).

The aim of this chapter is to provide with a brief overview of high performance sensorless induction motor drive. There exist two approaches to speed estimation for sensorless control of induction machine: non model based approach and model based approach. The non model based technique tracks a machine anisotropy: either saturation for flux estimation or rotor slotting for rotor position estimation, whereas the model based technique rely mostly on back emf voltage associated with fundamental component excitation of the machine. Model-based observers are considered very well adapted for state estimation and allow, in most cases, a stability proof and a methodology to tune observer gains. Among the observers the reduced order observers are more frequently implemented than the full order ones as they don't require heavy computations. Two sensorless vector control strategies using machine model-based estimation are presented in this chapter.

SPEED ESTIMATION

Rotor speed has been considered as a constant by many researchers in speed estimation problem (Schauder, 1992; Tajima & Hori, 1993; Peng & Fukao, 1994; Kim et al., 1994; Comnac et al., 2001; Ma & Gui, 2002; Du et al., 1995; Minami et al., 1991; Veleyez-Reyes & Verghese., 1992; Veleyez-Reyes et al., 1989). The idea is that the speed changes slowly compared to electrical variables. Adopting such an approach allowed speed estimation without requiring the knowledge of mechanical parameters of the drive system such as load torque, inertia etc. In (Schauder, 1992; Tajima & Hori, 1993; Peng & Fukao, 1994) speed was estimated using model reference adaptive system considering it as an unknown constant parameter. In (Kim et al., 1994; Comnac et al., 2001; Ma & Gui, 2002; Du et al., 1995) the speed was considered as an unknown constant state of the machine and extended kalman filter (EKF) was used to estimate it. Recursive least square estimation method was used in (Minami et al., 1991; Veleyez-Reyes & Verghese, 1992; Veleyez-Reyes et al., 1989) for speed estimation considering speed as an unknown constant parameter and found out the value of estimated speed that best fits the measured and calculated data in the dynamic equations of the motor.

In this section we present a sensorless vector control strategy using machine model-based speed estimation (Thongam & Ouhrouche, 2007). The proposed method does not require taking derivative of the measured signals unlike that of (Peng & Fukao, 1994; Minami et al., 1991; Veleyez-Reyes & Verghese, 1992; Veleyez-Reyes et al., 1989). The method is also simpler to implement than implementing EKF. In this method the model of the motor used for estimation is derived by introducing a new variable which is a function of rotor flux and speed assuming that rotor speed varies slowly in comparison to electrical states and hence its derivative can be conveniently equated to zero in the machine model used for estimation. The sensorless method presented here in this chapter is based on observing this new variable. A reduced order observer is implemented for estimating the new variable using which the rotor speed is estimated.

Induction Machine Model

The induction motor model in stationary stator reference frame α – β may be written in vector matrix form as

$$\frac{d\psi_r}{dt} = A_{11}\psi_r + A_{12}i_s \tag{1}$$

$$\frac{di_s}{dt} = A_{21}\psi_r + A_{22}i_s + A_{23}v_s \tag{2}$$

where

$A_{11} = -(R_r/L_r)I + \omega J$, $A_{12} = (L_m R_r / L_r)I$, $A_{21} = \frac{L_m}{\sigma L_s L_r}\{(R_r/L_r)I - \omega J\}$,

$A_{22} = -\{R_s/(\sigma L_s) + R_r L_m^2/(\sigma L_s L_r^2)\}I$, $A_{23} = 1/(\sigma L_s)I$, $I = \begin{bmatrix} 1 & 0 \\ 0 & 1 \end{bmatrix}$, $J = \begin{bmatrix} 0 & -1 \\ 1 & 0 \end{bmatrix}$, $\psi_r = [\psi_{r\alpha} \ \psi_{r\beta}]^T$ is the rotor flux, $i_s = [i_{s\alpha} \ i_{s\beta}]^T$ is the stator current, $v_s = [v_{s\alpha} \ v_{s\beta}]^T$ is the stator voltage and $\sigma = 1 - L_m^2/(L_s L_r)$ is the leakage coefficient.

Now, we introduce a new quantity into the motor model which when introduced will make the right hand side of conventional motor model given by equations (1) and (2) independent of the unknowns – the rotor flux and speed. Let's define the new quantity as

$$Z = -A_{11}\psi_r \tag{3}$$

A new motor model is obtained after introducing the new quantity as given below:

$$\frac{d\psi_r}{dt} = A_{12}i_s + A_{14}Z \tag{4}$$

$$\frac{di_s}{dt} = A_{22}i_s + A_{23}v_s + A_{24}Z \tag{5}$$

$$\frac{dZ}{dt} = A_{32}i_s + A_{34}Z \tag{6}$$

where

$A_{14} = -I$, $A_{24} = \{L_m/(\sigma L_s L_r)\}I$, $A_{32} = (L_m R_r^2/L_r^2)I - \omega(L_m R_r/L_r)J$ and $A_{34} = A_{11}$.

Observer Structure and Speed Estimation

The proposed speed estimation algorithm is based on observing the newly defined quantity which is a function of rotor flux and speed. Equation (5) and (6) are used for constructing a Gopinath's reduced order observer (Gopinath, 1971) for estimating the newly defined quantity. The observer is as given below

$$\frac{d\hat{Z}}{dt} = A_{32}i_s + A_{34}\hat{Z} + G\left(\frac{di_s}{dt} - \frac{d\hat{i}_s}{dt}\right) \quad (7)$$

where $G = \begin{bmatrix} g_1 & -g_2 \\ g_2 & g_1 \end{bmatrix}$ is the observer gain. Using equation (5) for $\frac{d\hat{i}_s}{dt}$ the observer equation

$$\frac{d\hat{Z}}{dt} = A_{32}i_s + A_{34}\hat{Z} + G\left(\frac{di_s}{dt} - A_{22}i_s - A_{23}v_s - A_{24}\hat{Z}\right) \quad (8)$$

The observer poles can be placed at the desired locations in the stable region of the complex plane by properly choosing the values of the elements of the G matrix. In order to avoid taking derivative of the stator current in the algorithm we introduce another new quantity

$$D = \hat{Z} - Gi_s \quad (9)$$

Finally, the observer is of the following form:

$$\frac{d}{dt}F = (A_{32} + A_{34}G - GA_{22} - GA_{24}G)i_s - GA_{23}v_s + (A_{34} - GA_{24})D \quad (10)$$

$$\hat{Z} = D + Gi_s \quad (11)$$

The block diagram of the Z observer is shown in Fig. 1.

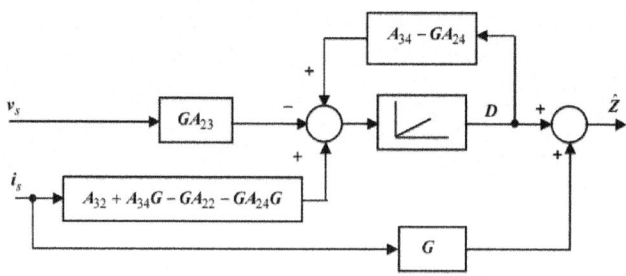

Figure 1: Block diagram of Z observer

Assuming no parameter variation and no speed error, the equation for error dynamics is given by

$$\frac{d}{dt}\tilde{Z} = \frac{d}{dt}(Z - \hat{Z}) = (A_{34} - A_{24}G)\tilde{Z} \qquad (12)$$

Eigenvalues of $(A_{34} - A_{24}G)$ are the observer poles which are as given below:

$$P_{obs1,2} = -\left(\frac{R_r}{L_r} + \frac{L_m}{\sigma L_s L_r}g_1\right) \pm j\left(\omega - \frac{L_m}{\sigma L_s L_r}g_2\right) \qquad (13)$$

The desired observer dynamics can be imposed by proper selection of observer gain G. Next, let's see how the rotor speed is computed. It can be seen that the observed quantity is a function of rotor flux and speed. Performing matrix multiplication of $\psi_r^T J$ with equation (3) we have

$$Z_\alpha \psi_{r\beta} - Z_\beta \psi_{r\alpha} = \left(\psi_{r\alpha}^2 + \psi_{r\beta}^2\right)\omega \qquad (14)$$

This is a simple equation which does not involve derivative or integration. To use it directly for speed computation we need to know the rotor flux; and as for Z_α and Z_β we can use the estimated values. The required flux is obtained from the reference. Rearranging the above equation we have the equation used for rotor speed computation as given by

$$\hat{\omega} = \frac{\hat{Z}_\alpha \psi_{r\beta}^* - \hat{Z}_\beta \psi_{r\alpha}^*}{\psi_{r\alpha}^{*2} + \psi_{r\beta}^{*2}} \qquad (15)$$

The coefficient matrices A_{32} and A_{34} in the observer equation are updated with the estimated values of rotor speed.

It is to be noted here that the model of the motor used in implementing the observer algorithm has been developed assuming that the derivative of the rotor speed is zero. It is valid to make such an assumption since the dynamics of rotor speed is much slower than that of electrical states. Moreover, such an assumption allows estimation without requiring the knowledge of mechanical quantities of the drive such as load torque, inertia etc.

Simulation Results

Simulation is carried out in order to validate the speed estimation algorithm presented. The block diagram of the sensorless indirect vector controlled induction motor drive incorporating the proposed speed estimator is shown in Fig. 2. The results of simulation are shown in Fig. 3 - Fig. 5.

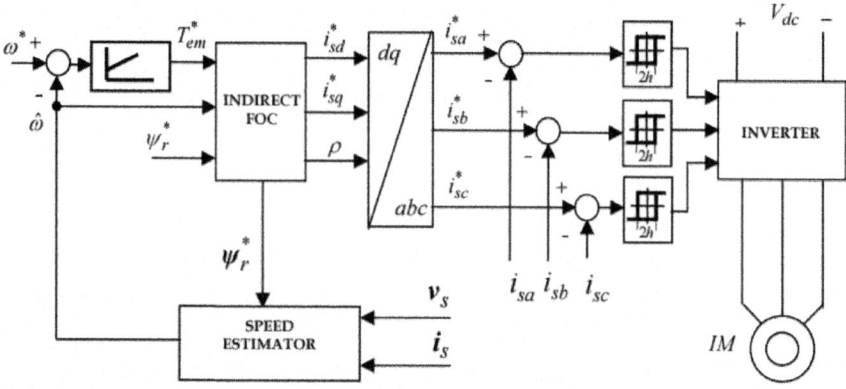

Figure 2: Sensorless indirect VC induction motor drive

Initially, the drive is run at no load. It is accelerated from rest to 150 rad/s at 0.15 sec. and then, the speed is reversed at 2.5 sec. The speed is reversed again at 5.5 sec. The speed of the motor (ω), estimated speed ($\hat{\omega}$) and reference speed (ω^*) are shown in Fig. 3 (a). Fig. 3 (b) shows speed estimation error ($\omega - \hat{\omega}$). The newly defined quantity (Z) and its estimated value ($\hat{Z}$) are shown in Fig. 3 (c) and its estimation error ($Z - \hat{Z}$) is shown in Fig. 3 (d).

The estimation algorithm and the drive response are then verified under loading and unloading conditions. The unloaded drive is started at 0.15 sec and full load is applied at 1 sec; then load is completely removed at 2 s. Later, after speed reversal, full load is applied at 4 sec and the load is completely removed at 5 sec. Fig. 4 shows the speed estimation result and response of the sensorless drive system.

Then, the sensorless induction motor drive is run under fully loaded condition at various operating speeds. The drive is started at full load at 0.15 s to 150 rad/s and the speed is reduced in steps in order to observe the response of the loaded drive at various speeds. Fig. 5 shows the estimation results and response of the loaded drive.

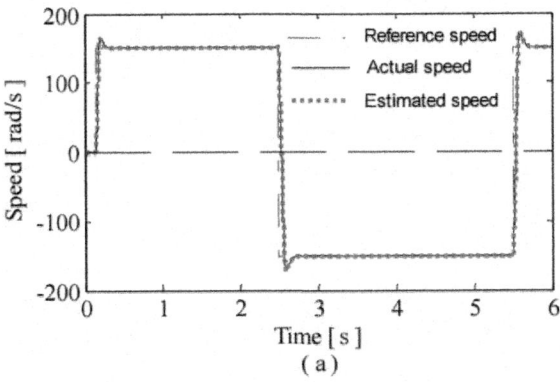

(a)

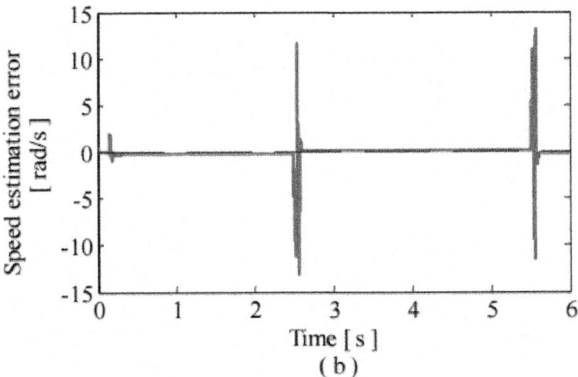

(b)

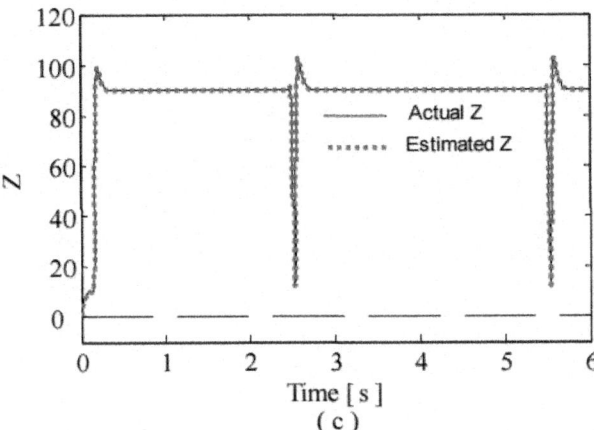

(c)

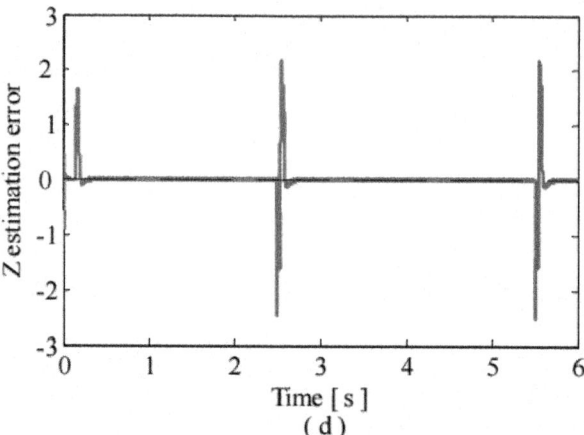

Figure 3: Acceleration and speed reversal at no load; (a) reference, actual and estimated speeds; (b) speed estimation error; (c) actual Z and estimated Z and (d) Z estimation error

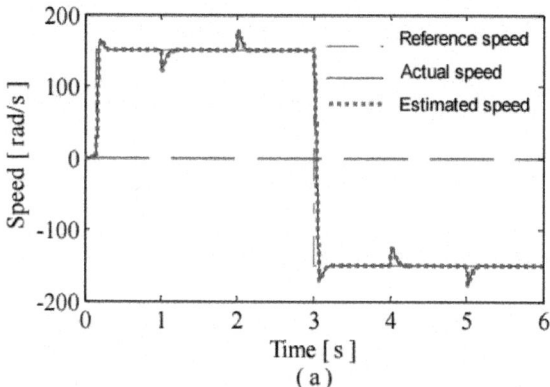

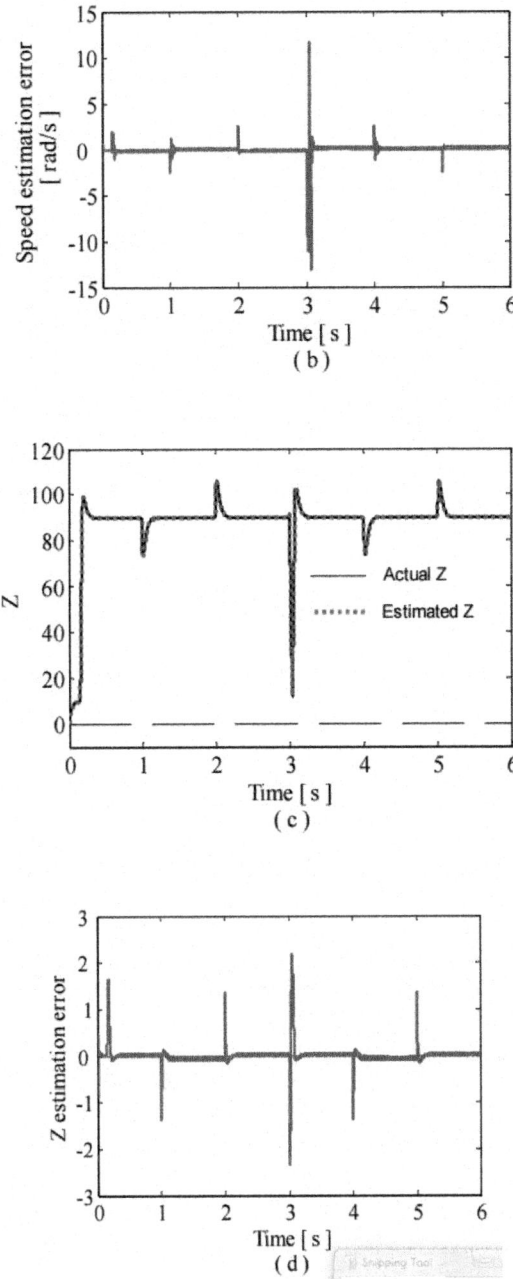

Figure 4: Application and removal of load; (a) reference, actual and estimated speeds; (b) speed estimation error; (c) actual Z and estimated Z and (d) Z estimation error

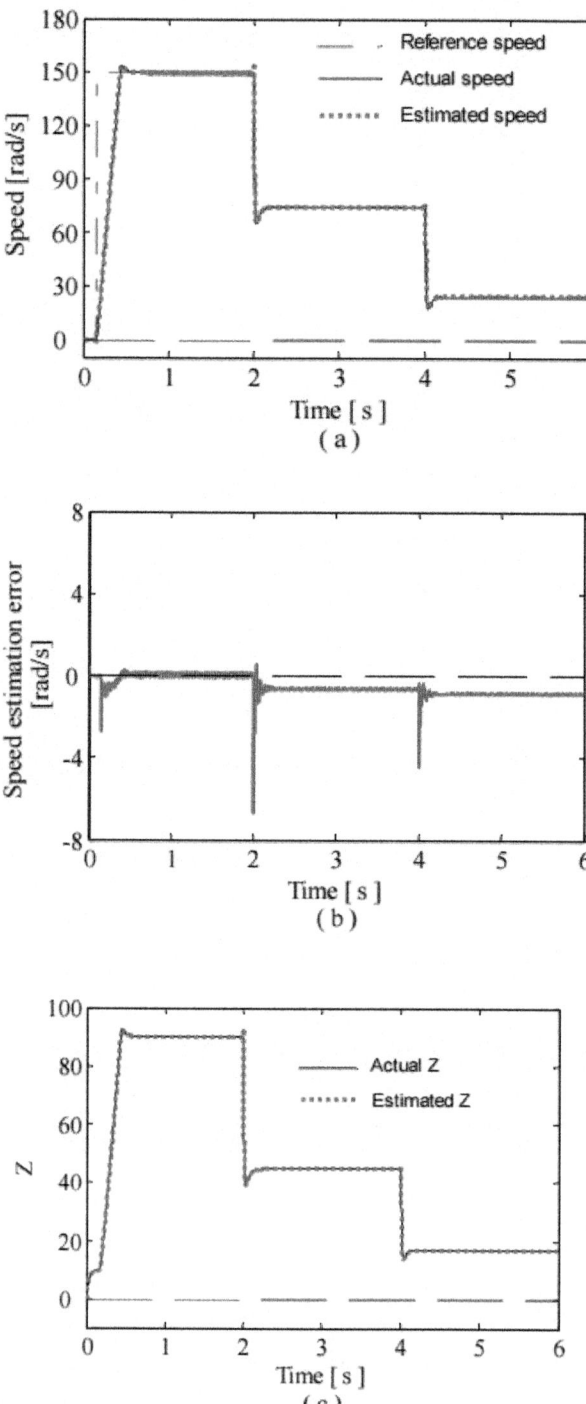

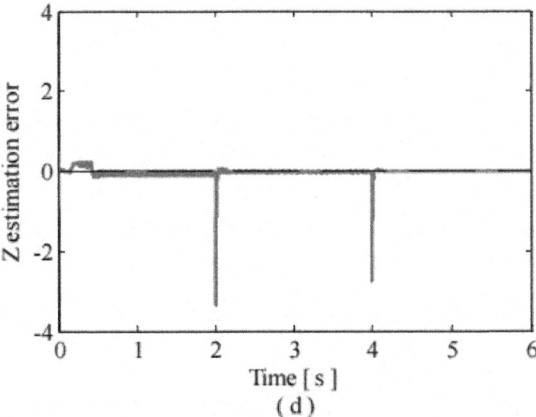

Figure 5: Operation at full load at various speeds; (a) reference, actual and estimated speeds; (b) speed estimation error; (c) actual Z and estimated Z and (d) Z estimation error

Improvement in Speed Estimation

It is observed that the estimation algorithm presented above gives good estimation accuracy under both dynamic and steady state conditions. However, it is found that the estimation accuracy decreases with decrease in speed. This is because of the fact that the estimation algorithm uses the command flux for speed estimation and not the actual rotor flux which however is little bit different from the command value. With the reduction in speed Z decreases and therefore the flux error is more prominently visible in the estimated speed. The problem is overcome in this section by using a rotor flux observer based on the voltage model of the machine along with the observer of the newly defined quantity. This allows accurate speed estimation in various operating ranges. The speed is computed using (15) after replacing the command flux by the estimated one. Further, due to the obvious advantages of dc current regulators over ac current regulators as regards its robustness, and load and operating point independence (Rowan & Kerkman, 1986) the control system uses dc current regulators. The rotor flux estimator and the control scheme are presented in the following subsections.

Flux estimation

The voltage model of induction motor is given by

$$\psi_r = \frac{L_r}{L_m}\left\{\int(v_s - R_s i_s)dt - \sigma L_s i_s\right\} \quad (16)$$

The rotor flux can be estimated using (16). However, the integration in (16) produces a problem of dc off-set and drift component in low speed region. Therefore, a first order low pass filter (LPF) is used instead of integration. The phase error in the low speed region produced due to LPF is approximately compensated by adding low pass filtered reference flux with the same time constant as above, and producing the estimated rotor flux (Ohtani et al., 1992). The estimator equation is given as

$$\hat{\psi}_r = \frac{L_r}{L_m}\left\{(v_s - R_s i_s)\frac{\tau}{1+\tau s} - \sigma L_s i_s \frac{\tau s}{1+\tau s}\right\} + \psi_r^* \frac{1}{1+\tau s} \quad (17)$$

where τ is the LPF time constant. The command rotor flux ψ_r^* in (17) is obtained as follows:

$$\psi_r^* = \begin{bmatrix}\psi_{r\alpha}^* \\ \psi_{r\beta}^*\end{bmatrix} = \begin{bmatrix}\psi_r^* \cos\rho^* \\ \psi_r^* \sin\rho^*\end{bmatrix} \quad (18)$$

where $\psi_r^* = L_m i_{sd}^*$ and ρ^*, the command rotor flux angle, is as given by

$$\rho^* = \int \omega_e^* dt \quad (19)$$

ω_e^*, the command rotor flux speed, is computed as given below

$$\omega_e^* = \omega_{sl}^* + \hat{\omega} \quad (20)$$

The command slip speed ω_{sl}^* is given by

$$\omega_{sl}^* = \frac{R_r i_{qs}^*}{L_r i_{ds}^*} \quad (21)$$

Speed estimation

The equation (15) after modification is used for speed computation. In place of reference flux, estimated flux is used for speed computation as given below

$$\hat{\omega} = \frac{\dot{Z}_\alpha \hat{\psi}_{r\beta} - \dot{Z}_\beta \hat{\psi}_{r\alpha}}{\hat{\psi}_{r\alpha}^2 + \hat{\psi}_{r\beta}^2} \quad (22)$$

Simulation Results

Simulation is carried out in order to verify the accuracy of the estimation algorithm and to see the response of the sensorless drive system. The block diagram of sensorless vector controlled induction motor drive incorporating the flux and speed estimators is shown in Fig. 6. First, the sensorless drive is run at no load at various speeds to verify the performance of the observer under no load condition. The drive is started at no-load and is run at various speeds by increasing it in steps to 10 rad/s, 50 rad/s, 100 rad/s and 150 rad/s at 0.3 sec, 1.5 sec, 3 sec and 4.5 sec respectively. The speed of the motor (ω), estimated speed ($\hat{\omega}$), reference speed (ω^*) and speed estimation error ($\omega - \hat{\omega}$) are shown in Fig. 7 (a). Fig. 7 (b) shows the actual Z, estimated Z and Z estimation error ($Z - \hat{Z}$).

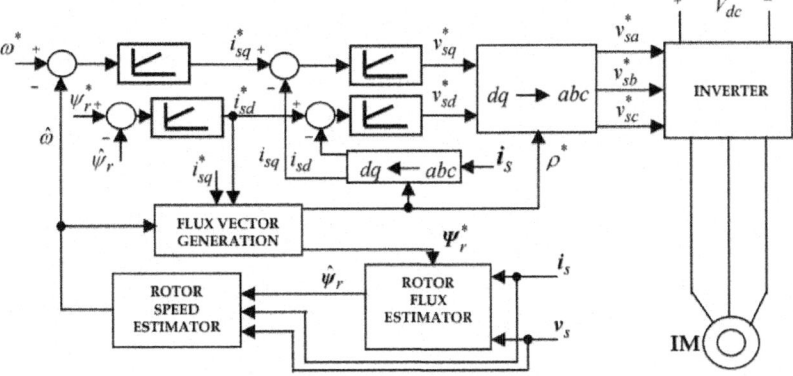

Figure 6: Sensorless vector controlled induction motor drive

Then, the performance of the estimator and drive response are verified on loading and unloading. The drive is started at no-load at 0.25 s to a speed of 150 rad/s and full load is applied at 1 sec and then the load is removed completely at 2 sec. Later, after speed reversal, full load is applied at 4 s and the load is completely removed at 5 sec. The response of the drive on application and removal of load is shown in Fig. 8.

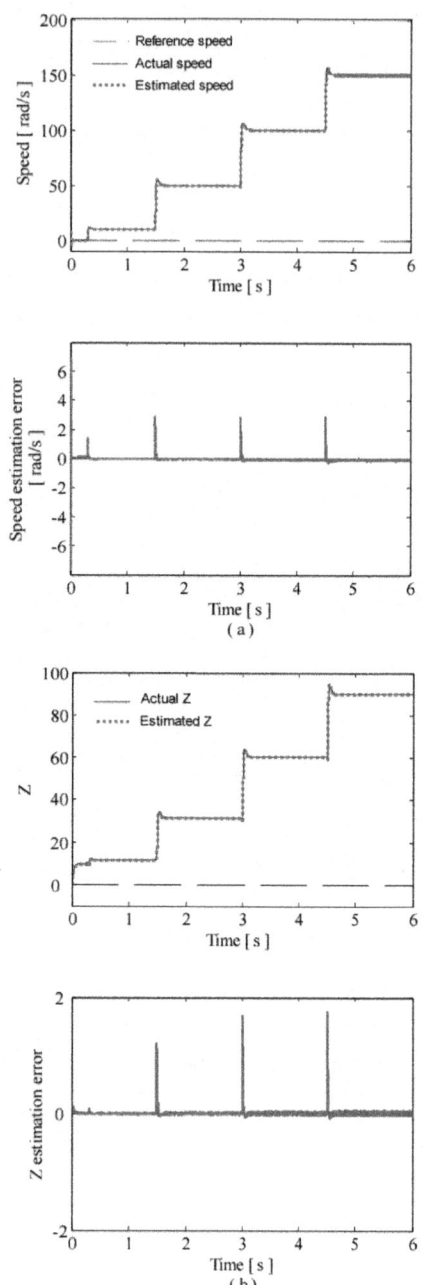

Figure 7: No load operation at various speeds; (a) reference, actual, estimated speeds, and speed estimation error; (b) actual Z, estimated Z, and Z estimation error

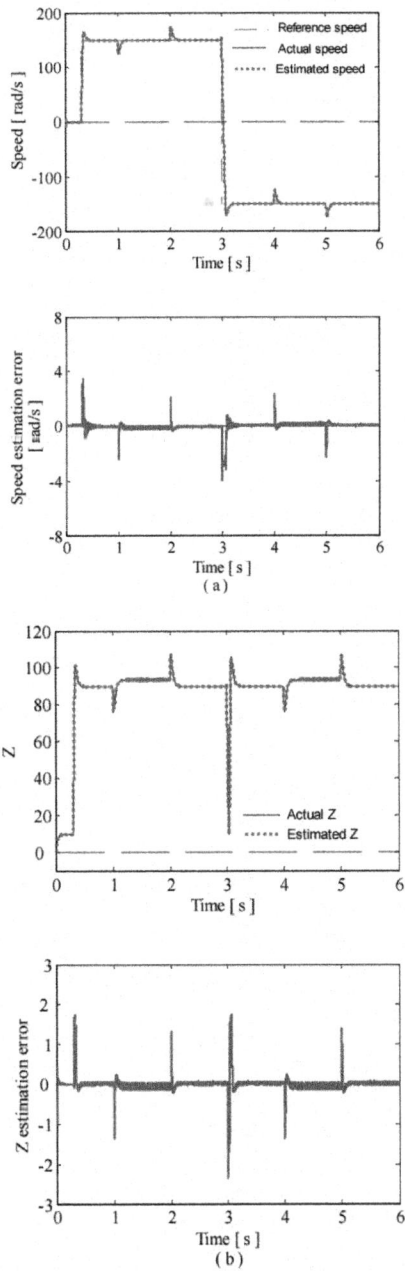

Figure 8: Application and removal of load; (a) reference, actual, estimated speeds, and speed estimation error; (b) actual Z, estimated Z, and Z estimation error.

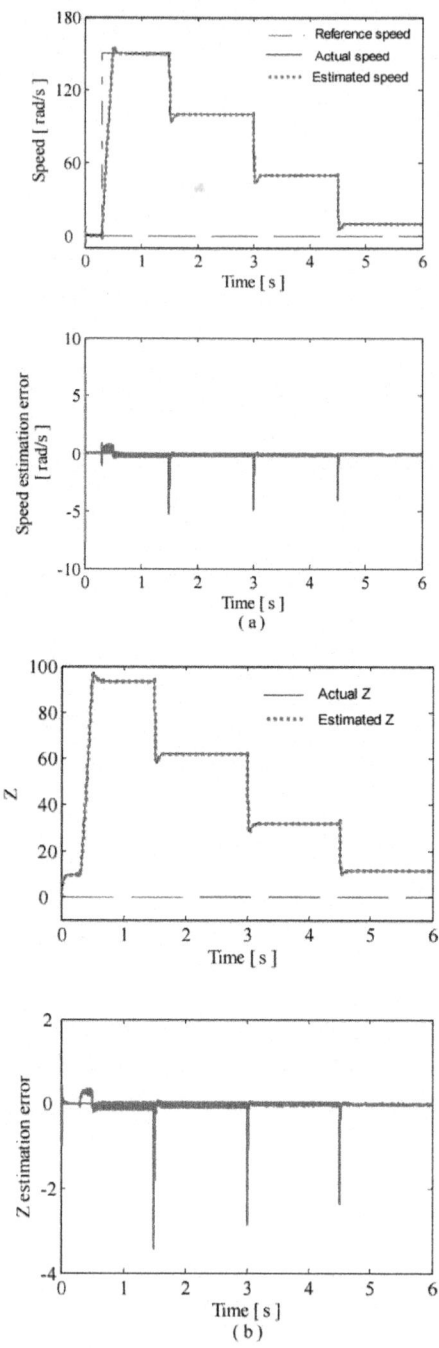

Figure 9: Full load operation at various speeds; (a) reference, actual, estimated speeds,

and speed estimation error; (b) actual Z, estimated Z, and Z estimation error.

The estimator performance and drive response are then verified under fully loaded condition of the drive at various operating speeds. The fully loaded machine is accelerated to 150 rad/s at 0.3 s, and then the speed is reduced in steps to 100 rad/s, 50 rad/s and 10 rad/s at 1.5 sec, 3 sec and 4.5 sec respectively. Fig. 9 shows the estimation results and response of the drive during the operation.

Good speed estimation accuracy was obtained under both dynamic and steady state conditions under various operating conditions and response of the VC induction motor drive incorporating the estimation algorithms was found to be good. The speed estimation algorithm presented in this section depends upon the knowledge of the rotor flux, whereas, the rotor flux estimator is independent of rotor speed and requiring only the measurable stator terminal quantities, the stator voltage and current.

FLUX AND SPEED ESTIMATION

Induction machines do not allow rotor flux to be easily measured. The current model and the voltage model are the traditional solutions, and their benefits and drawbacks are well known. Various observers for flux estimation were analyzed in the work by Verghese and Sanders (Verghese & Sanders, 1988) and Jansen and Lorenz (Jansen & Lorenz, 1994). Over the years several other have been presented, many of which include speed estimation (Tajima & Hori, 1993; Kim et al., 1994; Ohtani et al., 1992; Kubota et al., 1993; Sathiakumar, 2000; Yan et al., 2000).

Tajima & Hori (Tajima & Hori, 1993) proposed MRAS (Schauder, 1992) with novel pole allocation method for speed estimation while rotor flux estimation was done using Gopinath's observer. Extended Kalman Filter was used in (Kim et al., 1994) for estimating the rotor flux and speed using a full order model of the motor assuming that rotor speed is a constant. Ohtani et al (Ohtani et al., 1992) used the voltage model for flux estimation overcoming the problem associated with integrator and low pass filter while speed was obtained using a frequency controller. A speed adaptive flux observer was proposed in (Kubota et al., 1993) for estimating rotor flux and speed. Gopinath style reduced order observer was used in (Sathiakumar, 2000) for estimating the rotor flux while the speed was computed using an equation derived from the motor model. Yan et al (Yan et al., 2000) proposed a flux and speed estimator based on the sliding-mode control methodology.

In this section, we present a new flux estimation algorithm for speed sensorless rotor flux oriented controlled induction motor drive (Thongam & Ouhrouche, 2006). The proposed method is based on observing the variable

Z introduced in Section 2 which when introduced makes the right hand side of the conventional motor model independent of rotor flux and speed. Rotor flux estimation is achieved using an equation obtained after introduction of the newly defined quantity into the Blaschke equation or commonly known as the current model; while, speed is computed using a simple equation obtained using the new quantity Z.

Estimation of rotor flux and speed

The speed computation equation (22) obtained in section 2.4.2 requires the knowledge of rotor flux and Z. Here, we present a joint rotor flux and speed estimation algorithm. The block diagram of the proposed rotor flux and speed estimation algorithm is shown in Fig. 10.

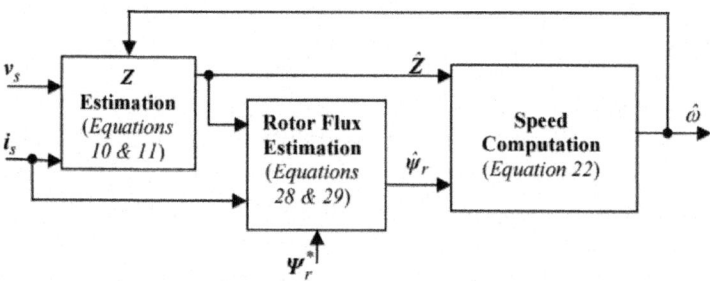

Figure 10: Rotor flux and speed estimator

Rotor flux may be obtained directly using equation (4) which is obtained after introducing the newly defined quantity Z into the Blaschke equation as

$$\hat{\psi}_r = \int (A_{12}i_s + A_{14}Z) dt \qquad (23)$$

However, rotor flux computation by pure integration suffers from dc offset and drift problems. To overcome the above problems a low pass filter is used instead of pure integrator and the phase error due to low pass filtering is approximately compensated by adding low pass filtered reference flux with the same time constant as used above (Ohtani et al., 1992). The equation of the proposed rotor flux estimator is given below

$$\hat{\psi}_r = \frac{\tau}{1+\tau s}(A_{12}i_s + A_{14}Z) + \frac{1}{1+\tau s}\psi_r^* \qquad (24)$$

where τ is the LPF time constant. The command rotor flux ψ_r^* is obtained as follows

$$\Psi_r^* = \begin{bmatrix} \Psi_{r\alpha}^* \\ \Psi_{r\beta}^* \end{bmatrix} = \Psi_r^* \begin{bmatrix} \cos\rho^* \\ \sin\rho^* \end{bmatrix} = L_m i_{sd}^* \begin{bmatrix} \cos\rho^* \\ \sin\rho^* \end{bmatrix} \quad (25)$$

The command rotor flux angle ρ^* is obtained by integrating the command rotor flux speed as given by

$$\rho^* = \int \omega_e^* dt = \int (\omega_{sl}^* + \hat{\omega}) dt \quad (26)$$

The command slip speed ω_{sl}^* is given by

$$\omega_{sl}^* = \frac{R_r i_{qs}^*}{L_r i_{ds}^*} \quad (27)$$

We know that the equation of the back emf is given by:

$$e = \frac{L_m}{L_r}\frac{d\psi_r}{dt} = \frac{L_m}{L_r}(A_{12}i_s + A_{14}Z) \quad (28)$$

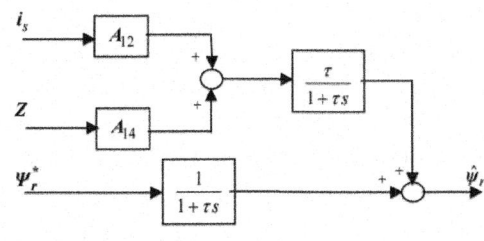

Figure 11: Rotor Flux Estimator

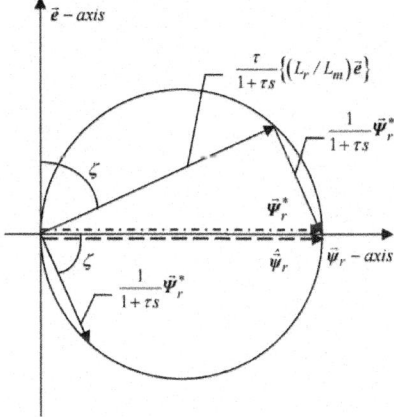

Figure 12: Obtaining estimated rotor flux

Now, equation (24) may also be written as

$$\hat{\psi}_r = \frac{\tau}{1+\tau s}\left(\frac{L_r}{L_m}e\right) + \frac{1}{1+\tau s}\psi_r^* \quad (29)$$

Block diagram of the rotor flux estimator is shown in Fig. 11. Fig.12 explains how estimated flux is obtained using equation (29).

Simulation Results

Simulation is carried out in order to validate the performance of the proposed flux and speed estimation algorithm. The proposed rotor flux and speed estimation algorithm is incorporated into a vector controlled induction motor drive. The block diagram of the sensorless vector controlled induction motor drive incorporating the proposed estimator is shown in Fig. 13. The sensorless drive system is run under various operating conditions.

First, acceleration and speed reversal at no load is performed. A speed command of 150 rad/s at 0.5 s is given to the drive system which was initially at rest, and then the speed is reversed at 3 s. The response of the drive is shown in Fig. 14. Fig. 14 (a) shows reference (ω^*), actual (ω), estimated $(\hat{\omega})$ speed, and speed estimation error $(\omega - \hat{\omega})$. The module of the actual $(|\Psi_r|)$, estimated $(|\hat{\Psi}_r|)$ rotor flux, and rotor flux estimation error $(|\Psi_r| - |\hat{\Psi}_r|)$ are shown in Fig. 14 (b). Fig. 14 (c) and (d) shows respectively the locus of the actual and estimated rotor fluxes.

The drive is then run at various speeds under no load condition. It is accelerated from rest to 10 rad/s at 0.5 s, then accelerated further to 50 rad/s, 100 rad/s and 150 rad/s at 1.5 s, 3 s and 4.5 s respectively. Fig. 15 shows the estimation of rotor flux and speed, and the response of the sensorless drive system.

Then, the drive is subjected to a slow change in reference speed profile (trapezoidal), the results of which are shown in Fig. 16.

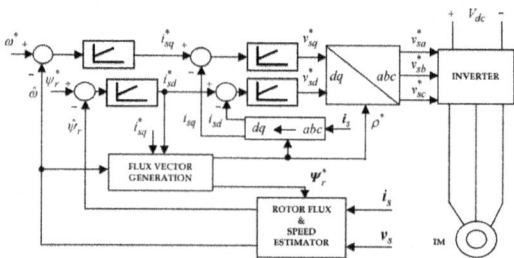

Figure 13: Sensorless vector controlled induction motor drive

Further, the performance of the estimator is verified under loaded conditions at various operating speeds. The fully loaded drive is accelerated to 150 rad/s at 0.5 s and then decelerated in steps to 100 rad/s, 50 rad/s and 10 rad/s at 1.5 s, 3 s and 4.5 s respectively. Fig. 17 shows the estimation results and response of the loaded drive system.

Then, we test the performance of the estimator on loading and unloading. The drive at rest is accelerated at no-load to 150 rad/s at 0.5 s and full load is applied at 1 s; we then remove the load completely at 2 s. Later, after speed reversal, full load is applied at 4 s, then, the load is removed completely at 5 s. Fig. 18 shows the estimation results and the response of the sensorless drive.

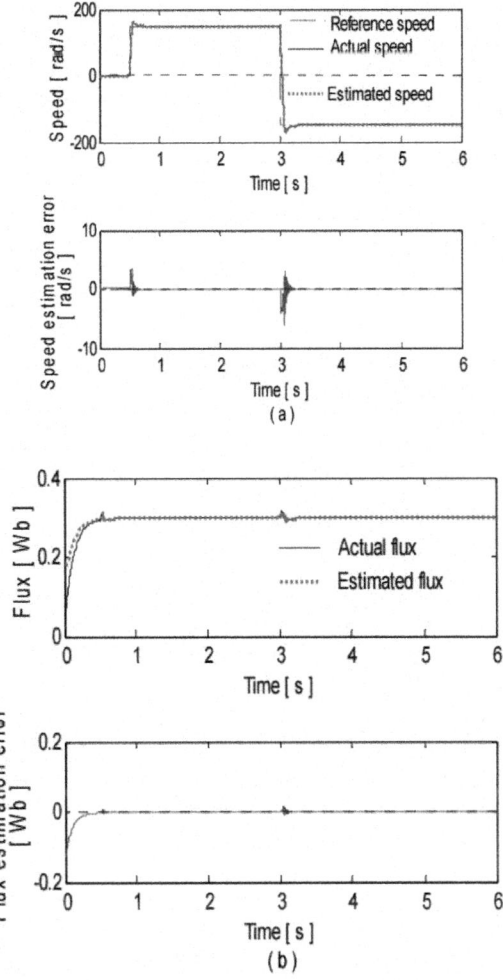

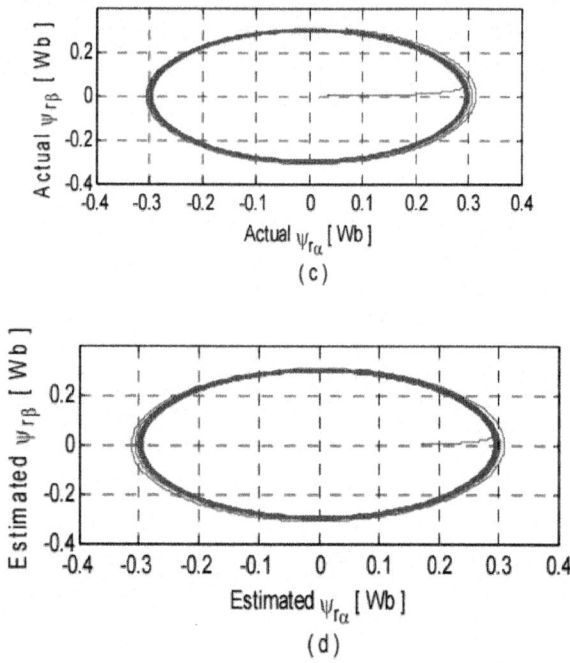

Figure 14: Acceleration and speed reversal of the sensorless drive at no-load

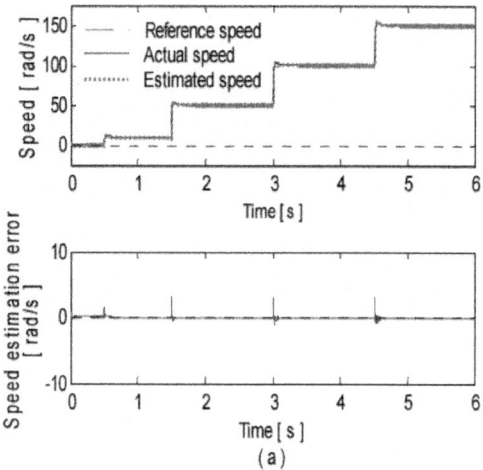

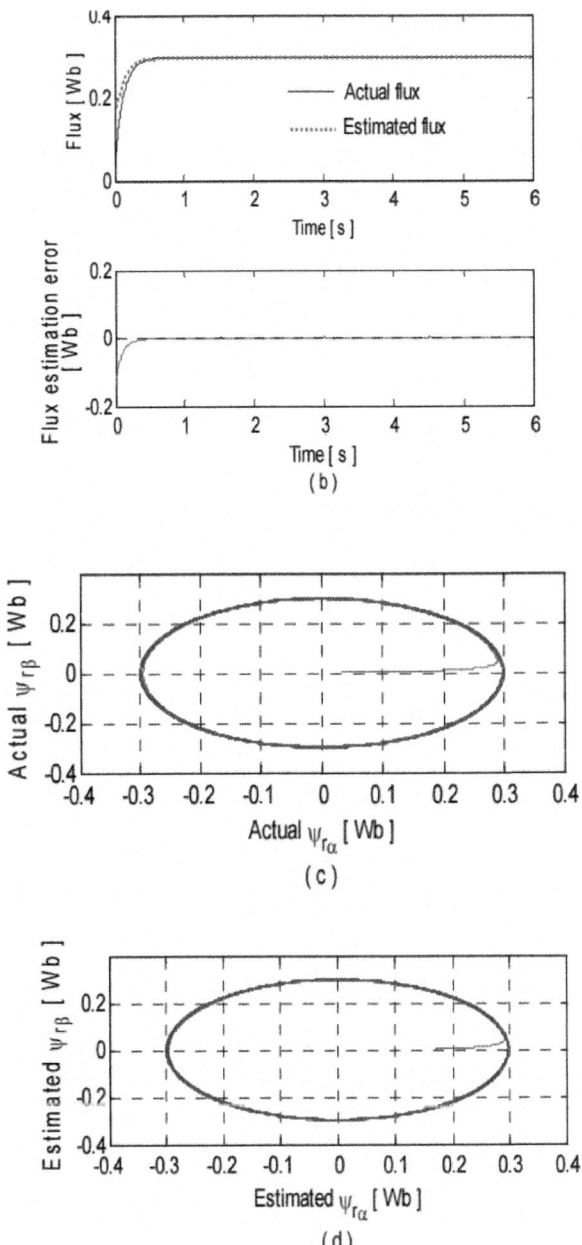

Figure 15: No-load operation of the sensorless drive with step increase in speeds

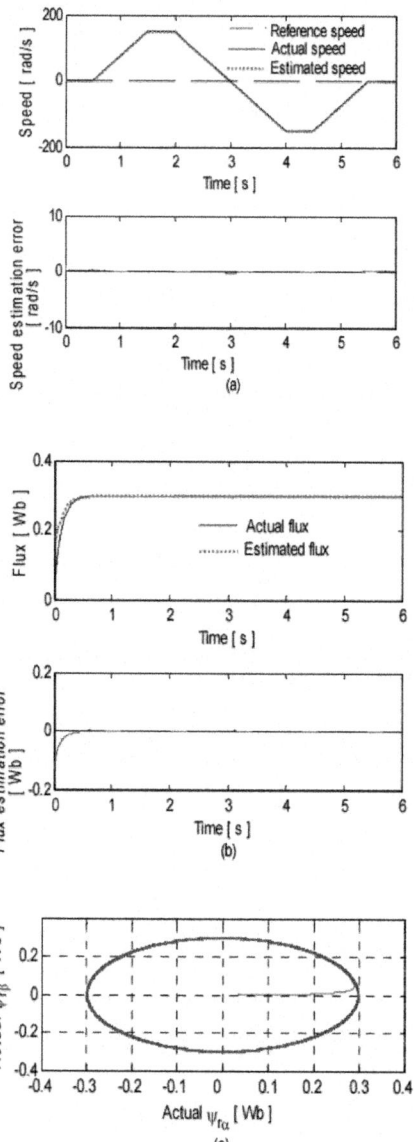

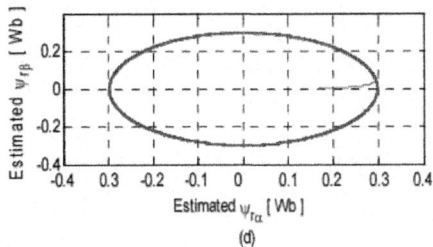

(d)

Figure 16: No-load operation of the sensorless drive with trapezoidal reference speed

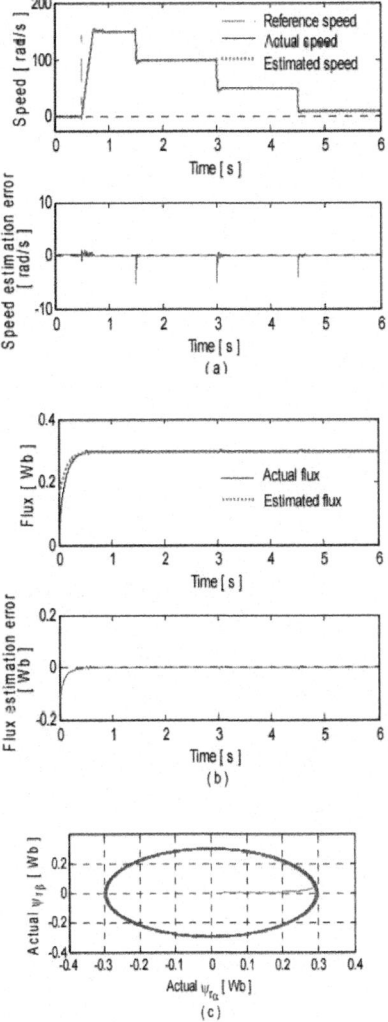

156 Electrical Machine Principles: A Handbook

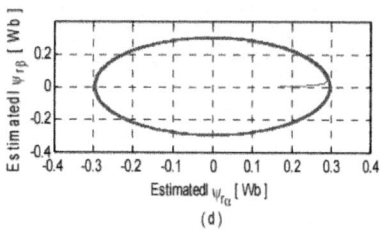

Figure 17: Operation of the sensorless drive at full load at various speeds

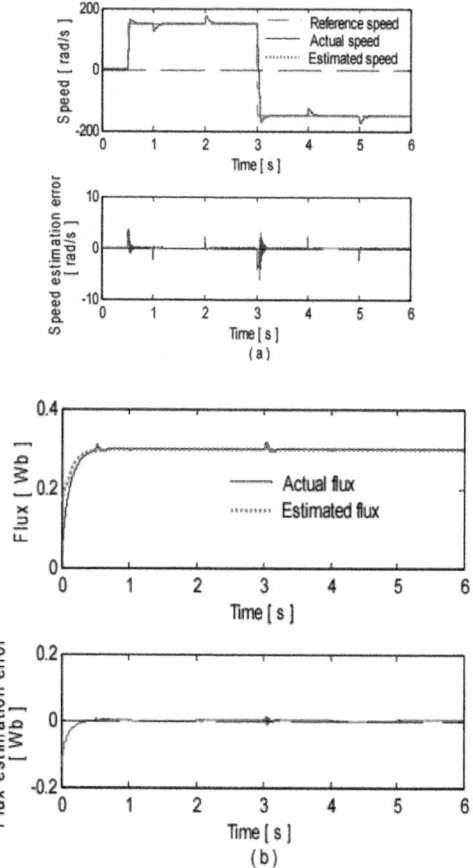

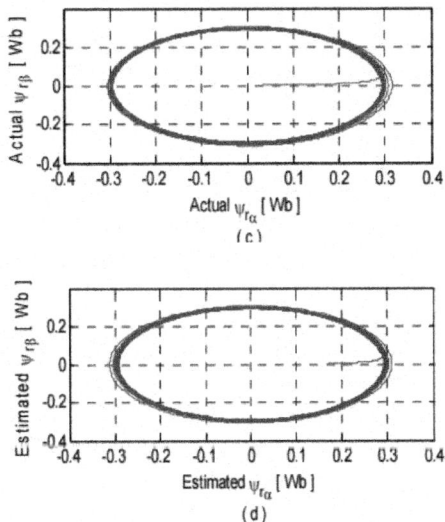

Figure 18: Drive response on application and removal of load

CONCLUSION AND FUTURE WORKS

In this chapter we have presented some methods of sensorless vector control of induction motor drive using machine model-based estimation. Sensorless vector control is an active research area and the treatment of the whole model based sensorless vector control will demand a book by itself.

First, a speed estimation algorithm in vector controlled induction motor drive has been presented. The proposed method is based on observing a newly defined quantity which is a function of rotor flux and speed. The algorithm uses command flux for speed computation. The problem of decrease in estimation accuracy with the decrease in speed was overcome using a flux observer based on voltage model of the machine along with the observer of the newly defined quantity, and satisfactory results were obtained.

Then, a joint rotor flux and speed estimation algorithm has been presented. The proposed method is based on a modified Blaschke equation and on observing the newly defined quantity mentioned above. Good estimation accuracy was obtained and the response of the sensorless vector controlled drive was found to be satisfactory.

The mathematical model of the motor used for implementing the estimation algorithm was derived with the assumption that the rotor speed dynamics is much slower than that of electrical states. Therefore, increase in estimation accuracy of the proposed algorithms will be observed with the increase in the size of the machine used. The machine model developed in this chapter may be used in future for machine parameter estimation. The newly defined quantity presented in this chapter contains rotor resistance information as well, in addition to that of rotor flux and speed. Therefore, future research efforts may be made towards developing rotor resistance estimation algorithm using the new machine model. Further, in the proposed algorithms rotor flux was necessary for speed estimation. Future research efforts may also be made towards developing a speed estimation algorithm for which the knowledge of rotor flux is not necessary.

REFERENCES

1. Abbondante, A. & Brennen, M. B. (1975). Variable speed induction motor drives use electronic slip calculator based on motor voltages and currents. *IEEE Trans. Ind. Appl,* Vol. 1A-11, No. 5, Sept/Oct, pp. 483-488.

2. Ben-Brahim, L. & Kudor, T. (1995). Implementation of an induction motor speed estimator using neural networks. *Proceedings of International Power Electronics Conference, IPEC 1995,* Yokohama, April, pp. 52-58.

3. Bodson, M.; Chiasson, J. & Novotnak, R. T. (1995). Nonlinear Speed Observer for High Performance Induction Motor Control. *IEEE Trans. Ind. Elec,* Vol. 42, No. 4, Aug. pp. 337-343.

4. Choy, I.; Kwon, S. H.; Lim J. & Hong, S. W. (1996). Robust Speed Estimation for Tacholess Induction Motor Drives. *IEEE Electronics Letters,* Vol. 32, No. 19, pp. 1836-1838.

5. Comnac V.; Cernat M.; Cotorogea, M. & Draghici, I. (2001). Sensorless Direct Torque and Stator Flux Control of Induction Machines Using an Extended Kalman Filter",

6. *Proceedings of IEEE Int. Conf. on Control Appl,* Mexico, Sept. 5-7, pp. 674-679.

7. Du T.; Vas, P. & Stronach, F. (1995). Design and Application of Extended Observers for Joint State and Parameter Estimation in High Performance AC Drives. *IEE Proc. Elec. Power Appl.,* Vol. 142, No. 2, pp. 71-78.

8. Fodor, D. ; Ionescu, F. ; Floricau, D. ; Six, J.P. ; Delarue, P. ; Diana, D. & Griva, G. (1995). Neural Networks Applied for Induction Motor Speed Sensorless Estimation. *Proceedings of the IEEE International Symposium on Industrial Electronics, ISIE' 95,* July 10-14, Athens, pp. 181-186.

9. Gopinath, B. (1971). On the Control of Linear Multiple Input-Output Systems. Bell System Technical Journal, Vol. 50, No. 3, March, pp. 1063-

1081.

10. Haghgoeian, F.; Ouhrouche, M. & Thongam, J. S. (2005). MRAS-based speed estimation for an induction motor sensorless drive using neural networks. *WSEAS Transactions on Systems,* Vol. 4, No. 12, December, pp. 2346-2352.

11. Jansen, P. L. & Lorenz, R. D. (1994). A physically insightful approach to the design and accuracy assessment of flux observers for field oriented induction machine drives. *IEEE Trans. Ind. App.,* Vol. 30, No. 1, Jan. / Feb., pp. 101-110.

12. Kim, S. H.; Park, T. S.; Yoo, J. Y. & Park, G. T. (2001). Speed-sensorless vector control of an induction motor using neural network speed estimation. *IEEE Trans. Ind. Elec,* Vol. 48, No. 3, June, pp. 609-614.

13. Kim, Y. R.; Sul S. K. & and Park, M. H. (1994). Speed sensorless vector control of induction motor using extended Kalman filter. *IEEE Trans. Ind. Appl.,* Vol. 30, No. 5, Sept/Oct, pp. 1225-1233.

14. Kubota, H.; Matsuse K. & Nakano, T. (1993). DSP-based speed adaptative flux observer of induction motor. *IEEE Trans. Ind. Appl,* Vol. 29, No. 2, March/April, pp. 344-348.

15. Liu, J. J.; Kung, I. C. & Chao, H. C. (2001). Speed estimation of induction motor using a non-linear identification technique. *Proc. Natl. Sci. Counc. ROC (A),* Vol. 25, No. 2, pp. 107-114

16. Ma, X. & Gui, Y. (2002). Extended Kalman filter for speed sensor-less DTC based on DSP. *Proc. of the 4th World Cong. on Intelligent Control and Automation,* Shanghai, China, June 10-14, pp. 119-122.

17. Minami, K.; Veley-Reyez, M.; Elten, D.; Verghese, G. C. & Filbert, D. (1991). Multi-stage speed and parameter estimation for induction machines. Proceedings of the *IEEE Power Electronics Specialists Conf.,* Boston, USA, pp. 596-604.

18. Ohtani, T.; Takada, N. & and Tanaka, K. (1992). Vector control of induction motor without shaft encoder. *IEEE Trans. Ind. Appl,* Vol. 28, No. 1, Jan/Feb, pp. 157-164.

19. Pappano, V.; Lyshevski, S. E. & Friedland, B. (1998). Identification of induction motor parameters. *Proceedings of the 37th IEEE Conf. on Decision and Control,* Tampa, Florida, USA, December 16-18, pp. 989-994.

20. Peng, F. Z. & Fukao, T. (1994). Robust speed identification for speed sensorless vector control of induction motors. *IEEE Trans. Ind. Appl,* Vol. 30, No. 5, Sept/Oct., pp. 1234-1240.

21. Rowan, T. M. & Kerkman, R. J. (1986). A new synchronous current regulator and an analysis of current-regulated PWM inverters. *IEEE Trans. Ind. Appl*, Vol. IA-22, No. 4, July/Aug., pp. 678-690.
22. Schauder, C. (1992). Adaptive speed identification for vector control of induction motors without rotational transducers. *IEEE Trans. Ind. Appl*, Vol. 28, No. 5, Sept./Oct., pp. 1054-1061.
23. Sathiakumar, S. (2000). Dynamic flux observer for induction motor speed control. *Proceedings of Australian Universities Power Engineering Conf. AUPEC 2000*, Brisbane, Australia, 24-27 Sept., pp. 108-113.
24. Simoes, M. G. & Bose, B. K. (1995). Neural network based estimation of feedback signals for a vector controlled induction motor drive. *IEEE Trans. Ind. Appl.*, Vol. 31, May/June, pp. 620-629.
25. Tajima, H. & Hori, Y. (1993). Speed sensorless field-orientation control of the induction machine. *IEEE Trans. Ind. Appl.*, Vol. 29, No. 1, pp. 175-180.
26. Thongam, J. S. & Thoudam, V. P. S. (2004). Stator flux based speed estimation of induction motor drive using EKF. *IETE Journal of Research*, India, Vol. 50, No. 3. May-June, pp 191-197.
27. Thongam, J. S. & Ouhrouche, M. (2006). Flux estimation for speed sensorless rotor flux oriented controlled induction motor drive. *WSEAS Transactions on Systems*, Vol. 5, No. 1, Jan., pp. 63-69.
28. Thongam, J. S. & Ouhrouche, M. (2007). A novel speed estimation algorithm in indirect vector controlled induction motor drive. *International Journal of Power and Energy Systems*, Vol. 27, No. 3, 2007, pp. 294-298.
29. Toqeer, R. S. & Bayindir, N. S. (2003). Speed estimation of an induction motor using Elman neural network. *Neuro Computing*, Volume 55, Issues 3-4, October, pp. 727- 730.
30. Velez-Reyes, M.; Minami, K. & Verghese, G. C. (1989). Recursive speed and parameter estimation for induction machines", *IEEE/IAS Ann. Meet. Conf. Rec.,* San Diego, pp. 607-611.
31. Veleyez-Reyes, M. & Verghese, G. C. (1992). Decomposed algorithms for speed and parameter estimation in induction machines. *IFAC Symposium on Nonlinear Control System Design,* Bordeaux, France, pp. 77-82.
32. Verghese, G. C. & Sanders, S. R. (1988). Observers for flux estimation in induction machines. *IEEE Trans. Ind. Elec*, Vol. 35, No. 1, Feb., pp. 85-94.
33. Yan, Z.; Jin C. & Utkin, V. I. (2000). Sensorless sliding-mode control of induction motors. *IEEE Trans. Ind. Elec*, Vol. 47, No. 6, Dec., pp. 1286-1297

Chapter 8

FEEDBACK LINEARIZATION OF SPEED-SENSORLESS INDUCTION MOTOR CONTROL WITH TORQUE COMPENSATION

Cristiane Cauduro Gastaldini[1], Rodrigo Zelir Azzolin[2],
Rodrigo Padilha Vieira[3], and Hilton Abílio Gründling[4]

[1,2,3,4]Federal University of Santa Maria
[2]Federal University of Rio Grande
[3]Federal University of Pampa Brazil

INTRODUCTION

This chapter addresses the problem of controlling a three-phase Induction Motor (IM) without mechanical sensor (i.e. speed, position or torque measurements). The elimination of the mechanical sensor is an important advent in the field of low and medium IM servomechanism; such as belt conveyors, cranes, electric vehicles, pumps, fans, etc. The absence of this sensor (speed, position or torque) reduces cost and size, and increases reliability of the overall system. Furthermore, these sensors are often difficult to install in certain applications and are susceptible to electromagnetic interference. In fact, sensorless servomechanism may substitute a measure value by an estimated one without deteriorating the drive dynamic performance especially under uncertain load torque.

Many approaches for IM sensorless servomechanism have been proposed in the literature is related to vector-controlled methodologies. One of the proposed nonlinear control methodologies is based on Feedback Linearization Control (FLC), as first introduced by (Marino et al., 1990). FLC provides rotor speed regulation, rotor flux amplitude decoupling and torque compensation. Although the strategy presented by (Marino et al., 1990) was not a sensorless control strategy, fundamental principles of FLC follow servomechanism design of sensorless control strategies, such as (Gastaldini & Grundling, 2009; Marino et al., 2004; Montanari et al., 2007; 2006).

The purpose of this chapter is to present the development of two FLC control strategies in the presence of torque disturbance or load variation, especially under low rotor speed conditions. Both control strategies are easily implemented in fixed point DSP, such as TMS320F2812 used on real time experiments and can be easily reproduced in the industry. Furthermore, an analysis comparing the implementation and the limitation of both strategies is presented. In order to implement the control law, these algorithms made use of only two-phase IM stator currents measurement. The values of rotor speed and load torque states used in the control algorithm are estimated using a Model Reference Adaptive System (MRAS) (Peng & Fukao, 1994) and a Kalman filter (Cardoso & Gründling, 2009), respectively.

This chapter is organized as follows: Section 2 presents the fifth-order IM mathematical model. Section 3 introduces the feedback linearization modelling of IM control. A simplified FLC control strategy is described in Section 4. The proposed methods for speed and torque estimation, MRAS and Kalman filter algorithms, respectively, are developed in Sections 5 and 6. State variable filter is used to obtain derivative signals necessary for implementation of the control algorithm, and this is presented in section 7. Digital implementation in fixed point DSP TMS320F2812 and real time experimental results are given in Section 8. Finally, Section 9 presents the conclusions.

INDUCTION MOTOR MATHEMATICAL MODEL

A three-phase N pole pair induction motor is expressed in an equivalent two-phase model in an arbitrary rotating reference frame (q-d), according to (Krause, 1986) and (Leonhard, 1996) according to the fifth-order model, as

$$\frac{d}{dt}I_{qs} = -a_{12}I_{qs} - \omega_s I_{ds} + a_{13}a_{11}\lambda_{qr} - a_{13}N\omega\lambda_{dr} + a_{14}V_{qs} \quad (1)$$

$$\frac{d}{dt}I_{ds} = -a_{12}I_{ds} + \omega_s I_{qs} + a_{13}a_{11}\lambda_{dr} + a_{13}N\omega\lambda_{qr} + a_{14}V_{ds} \quad (2)$$

$$\frac{d}{dt}\lambda_{qr} = -a_{11}\lambda_{qr} - (\omega_s - N\omega)\lambda_{dr} + a_{11}L_m I_{qs} \quad (3)$$

$$\frac{d}{dt}\lambda_{dr} = -a_{11}\lambda_{dr} + (\omega_s - N\omega)\lambda_{qr} + a_{11}L_m I_{ds} \quad (4)$$

$$\frac{d}{dt}\omega = \mu \cdot (\lambda_{dr}I_{qs} - \lambda_{qr}I_{ds}) - \frac{B}{J}\omega - \frac{T_L}{J} \quad (5)$$

$$T_e = \mu \cdot J \cdot (\lambda_{dr}I_{qs} - \lambda_{qr}I_{ds}) \quad (6)$$

In equations (1)-(6): $\mathbf{I_s} = (I_{qs}, I_{ds})$, $\mathbf{\lambda_r} = (\lambda_{qr}, \lambda_{dr})$ and $\mathbf{V_s} = (V_{qs}, V_{ds})$ denote stator current, rotor flux and stator voltage vectors, where subscripts d and q stand for vector components in (q-d) reference frame; ω is the rotor speed, the load torque T_L, electric torque T_e and ω_s is the stationary speed, θ0 is the angular position of the (q-d) reference frame with respect to a fixed stator reference frame (α-β), where physical variables are defined. Transformed variables related to three-phase (RST) system are given by xαβ =

$$x_{\alpha\beta} = K \cdot x_{RST} \tag{7}$$

Let
$$x_{qd} = e^{j\theta_0} x_{\alpha\beta} \tag{8}$$

with

$$e^{j\theta_0} = \begin{bmatrix} \cos\theta_0 & -\sin\theta_0 \\ \sin\theta_0 & \cos\theta_0 \end{bmatrix} \text{ and } K = \sqrt{\frac{2}{3}} \begin{bmatrix} 1 & -\frac{1}{2} & -\frac{1}{2} \\ 0 & -\frac{\sqrt{3}}{2} & \frac{\sqrt{3}}{2} \end{bmatrix}.$$

x_{qd} and $x_{\alpha\beta}$ stand for two-dimensional voltage flux and stator current vector, respectively on (q-d) and (α-β) reference frame.

The relations between mechanical and electrical parameters in the above equations are

$$a_0 \triangleq L_s L_r - L_m^2, \; a_{11} \triangleq \frac{R_r}{L_r}, \; a_{12} \triangleq \left(\frac{L_s L_r}{a_0} \frac{R_s}{L_s} + \frac{L_m^2}{a_0} a_{11} \right), \; a_{13} \triangleq \frac{L_m}{a_0}, \; a_{14} \triangleq \frac{L_r}{a_0} \text{ and } \mu \triangleq \frac{NL_m}{JL_r};$$

where R_s, R_r, L_s and L_r are the stator/rotor resistances and inductances, L_m is the magnetizing inductance, J is the rotor inertia, B is the viscous coefficient and N is the number of pole pairs. In the control design, the viscous coefficient of (5) is considered to be approximately zero, i.e. B ≈ 0.

FEEDBACK LINEARIZATION CONTROL

The feedback Linearization Control (FLC) general specifications are two outputs - rotor speed and rotor flux modulus, as

$$y_1 = \begin{bmatrix} \omega & \sqrt{\lambda_{qr}^2 + \lambda_{dr}^2} \end{bmatrix}^T \triangleq \begin{bmatrix} \omega & |\lambda_r| \end{bmatrix}^T \tag{9}$$

which is controlled by two-dimensional stator voltage vector Vs, on the basis of measured variables vector y2 = Is. The development concept of this control strategy is completely described in (Marino et al., 1990) and it will be omitted here. Following the concept of indirect field orientation developed by Blaschke, (Krause, 1986) and (Leonhard, 1996), the purpose of FLC control is to align rotor flux vector with the d-axis reference frame, i.e.

$$\lambda_{dr} = |\lambda_{\mathbf{r}}| \qquad \lambda_{qr} = 0 \qquad (10)$$

The condition expressed in (10) guarantees the exact decoupling of flux dynamics of (1)-(4) from the speed dynamics. Once rotor flux is not directly measured, only asymptotic field orientation is possible, according to (Marino et al., 1990) and (Peresada & Tonielli, 2000), then

$$\lim_{t\to\infty} \lambda_{dr} = |\lambda_{\mathbf{r}}| \qquad \lim_{t\to\infty} \lambda_{qr} = 0 \qquad (11)$$

It is defined $y_1^* = [\ \omega_{ref}\ \ \lambda_r^*\ ]^T$, where ω_{ref} and λ_r^* are reference trajectories of rotor speed and rotor flux. The speed tracking, flux regulation control problem under speed sensorless conditions is formulated considering IM model (1)-(5) under the following conditions: (a) Stator currents are measurable;

(a) Stator currents are measurable;
(b) Motor parameters are known and considered constant;
(c) Load torque is estimated and it is applied after motor flux excitation;
(d) Initial conditions of IM state variables are known;

(e λ_r^* is the flux constant reference value and estimated speed $\hat{\omega}$ and reference speed ω_{ref} are the smooth reference bounded speed signals

FLC equations are developed considering the fifth-order IM model under the assumption that estimated speed tracks real speed, and therefore it is acceptable to replace measured speed with estimated speed $(i.e.\ \hat{\omega}_k \approx \omega)$. In addition, the torque value is estimated using a Kalman filter. Fig. 1 presents the block diagram of FLC Control.

Flux Controller

From the decoupling properties of field oriented transformation (10), the control objective of the flux controller is to generate a flux vector aligned with the d-axis to guarantee induction motor magnetization.

Then, substituting (10) in (4)

$$i_{ds}^* = \left(a_{11}\lambda_r^* + \frac{d}{dt}|\lambda_{\mathbf{r}}|\right)\frac{1}{a_{11}L_m}$$

$$(12)$$

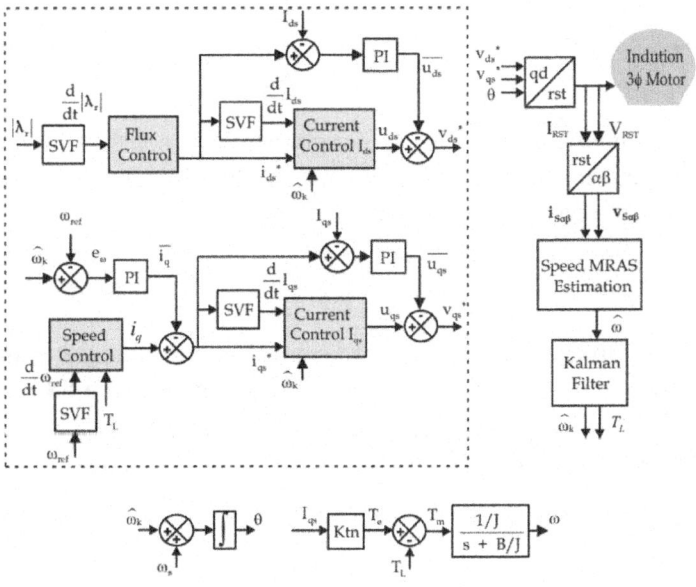

Figure 1: Feedback Linearization Control proposed

The rotor flux $|\lambda_r|$ is estimated by a model derived from the induction motor mathematical model, (3) and (4), that makes use of measured stator currents I_{qs}, I_{ds} and estimated speed $\hat{\omega}$ variables.

$$\frac{d}{dt}\lambda_r = -a_{11}\lambda_r - j(\omega_s - N\omega)\lambda_r + a_{11}L_m I_s \tag{13}$$

where the stationary speed is $\omega_s = N\hat{\omega} + \dfrac{a_{11}L_m}{\lambda^*}i_{qs}^*$. The digital implementation of the flux controller is made using Euler discretization and the derivative rotor flux signal is obtained by a state variable filter (SVF).

Speed Controller

The speed control algorithm uses the same strategy adopted for the flux subsystem and it is computed from (5), as

$$i_q = \frac{1}{\mu\lambda_r^*}\left(\frac{\hat{T}_L}{J} + \frac{d}{dt}\omega_{ref}\right) \tag{14}$$

To compensate for speed error between estimated speed and reference speed, (i.e. $e_\omega = \hat{\omega} - \omega_{ref}$), a proportional integral compensation is proposed, as follows

$$\overline{i_q} = \left(k_{p\_iq} + \frac{k_{i\_iq}}{s}\right) e_\omega \tag{15}$$

These gains values (k_{p_iq}, k_{i_iq}) are determined considering an induction motor mechanical model. The reference quadrature component stator speed current is derived from (14)-(15), as

$$i_{qs}^* = i_q - \overline{i_q} \tag{16}$$

In DSP implementation, the speed controller is discretized using the Euler method and the rotor speed derivative (14) is computed by a SVF.

Currents controller

From (1) and (2), the currents controller is obtained, as

$$u_{qs} = \frac{1}{a_{14}} \left(a_{12} i_{qs}^* + \omega_s i_{ds}^* + a_{13} \lambda_r^* N \left(\omega_{ref} + e_\omega\right) + \frac{d}{dt} I_{qs} \right) \tag{17}$$

and

$$u_{ds} = \frac{1}{a_{14}} \left(a_{12} i_{ds}^* + \omega_s i_{qs}^* + a_{11} a_{13} \lambda_r + \frac{d}{dt} I_{ds} \right) \tag{18}$$

where proportional integral gains of the current error

$$\overline{u}_{qs} = \left(k_{pv} + \frac{k_{iv}}{s}\right) \tilde{i}_{qs} \tag{19}$$

and

$$\overline{u}_{ds} = \left(k_{pv} + \frac{k_{iv}}{s}\right) \tilde{i}_{ds} \tag{20}$$

in which $\tilde{i}_{qs} = I_{qs} - i_{qs}^*$ and $\tilde{i}_{ds} = I_{ds} - i_{ds}^*$.

These gains (k_{pv}, k_{iv}) are determined considering a simplified induction motor electrical model, which is obtained by load and locked rotor test. Hence, current controllers are expressed as

$$v_{qs}^* = u_{qs} - \overline{u}_{qs} \tag{21}$$

$$v_{ds}^* = u_{ds} - \overline{u}_{ds} \qquad (22)$$

In DSP, currents controller are digitally implemented using discretized equation (17)-(22) based on the Euler method, and the stator current derivative is obtained by SVF using stator currents measures.

In order to reduce the number of computation requirements, a simplified feedback linearization control scheme is proposed. In this control scheme, one part of the current controller (6)-(7) is suppressed and only a proportional integral controller is used. This modification minimizes the influence of parameters variation in the control system.

Fig. 2 presents the block diagram of the Simplified FLC proposed. The currents controller of simplified FLC are defined as

$$v_{qs}^* = \left(k_{pv} + \frac{k_{iv}}{s}\right)\tilde{i}_{qs} \qquad (23)$$

$$v_{ds}^* = \left(k_{pv} + \frac{k_{iv}}{s}\right)\tilde{i}_{ds} \qquad (24)$$

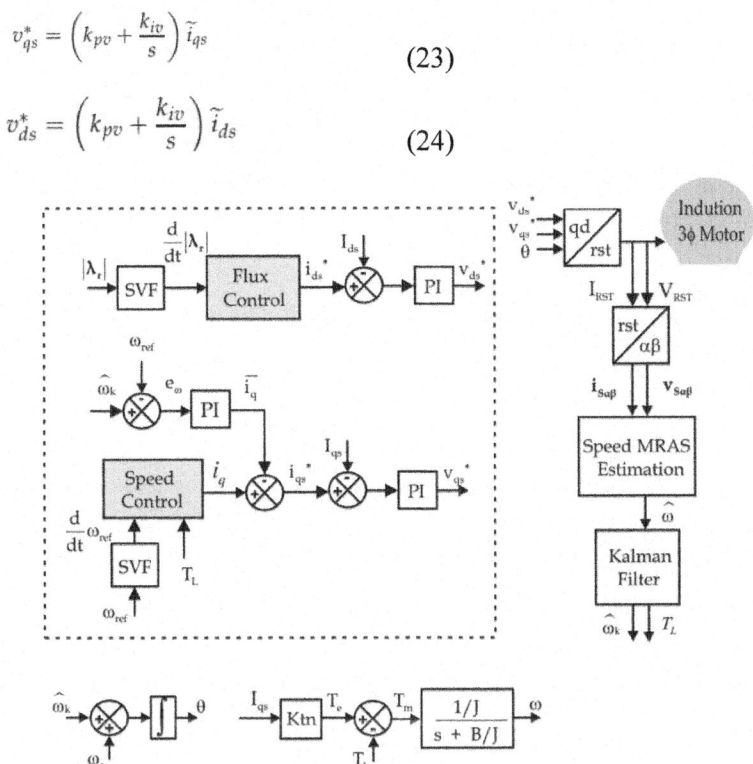

Figure 2: Proposed Simplified Feedback Linearization Control

Flux and Speed Controller are computed exactly as in the previous scheme, as (12) and (14)-(16).

SPEED ESTIMATION - MRAS ALGORITHM

A squirrel-cage three-phase induction motor model expressed in a stationary frame can be modelled using complex stator and rotor voltage as in (Peng & Fukao, 1994)

$$\mathbf{v_s} = R_s \mathbf{i_s} + L_s \frac{d}{dt}\mathbf{i_s} + L_m \frac{d}{dt}\mathbf{i_r} \quad (25)$$

for squirrel-cage IM $v_r = 0$

$$0 = R_r \mathbf{i_r} - jN\omega L_r \mathbf{i_r} - jN\omega L_m \mathbf{i_s} + L_r \frac{d}{dt}\mathbf{i_r} + L_m \frac{d}{dt}\mathbf{i_s} \quad (26)$$

The voltage and the current space vectors are given as $\bar{x} = x_\alpha + jx_\beta$, $\bar{x} \in \{\mathbf{v_s}, \mathbf{i_s}, \mathbf{i_r}\}$, relative to the transformed variables present in (7). The induction motor magnetizing current is expressed by

$$\mathbf{i_m} = \frac{L_r}{L_m}\mathbf{i_r} + \mathbf{i_s} \quad (27)$$

Two independent observers are derived to estimate the components of the counter-electromotive vectors.

$$\hat{\mathbf{e}}_m = \frac{L_m^2}{L_r}\mathbf{i_m} = \frac{L_m^2}{L_r}\left(\omega \mathbf{i_m} - \frac{1}{T_r}\mathbf{i_m} + \frac{1}{T_r}\mathbf{i_s}\right) \quad (28)$$

$$\mathbf{e}_m = \mathbf{v_s} - R_s \mathbf{i_s} - \sigma L_s \frac{d}{dt}\mathbf{i_s} \quad (29)$$

where $\sigma = 1 - \frac{L_m}{L_s L_r}$. The instantaneous reactive power maintains the magnetizing current, and its value is defined by cross product of the counter-electromotive and stator current vector

$$\mathbf{q}_m = \mathbf{i_s} \otimes \mathbf{e}_m \quad (30)$$

Substituting (28) and (29) for e_m in (30) and noting that is $\mathbf{i_s} \otimes \mathbf{i_s} = 0$, which gives

$$\mathbf{q}_m = \mathbf{i_s} \otimes \left(\mathbf{v_s} - \sigma L_s \frac{d}{dt}\mathbf{i_s}\right) \quad (31)$$

and

$$\hat{q}_m = \frac{L_m^2}{L_r}\left((i_m \odot i_s)\omega + \frac{1}{T_r}(i_m \otimes i_s)\right) \tag{32}$$

Then, q_m is the reference model of reactive power and $\hat{q}_m$ is the adjustable model. The estimated speed is produced by the proportional integral adaptation mechanism error of both models, and an MRAS system can be drawn as in Fig.3

This algorithm is customary for speed estimation and simple to implement in fixed point DSP, such as in (Gastaldini & Grundling, 2009; Orlowska-Kowalska & Dybkowski, 2010; Vieira et al., 2009).

The SVF blocks are state variable filters and are explained in greater detail in Section 7. These filters compute derivative signals and are applied in voltage signals to avoid addition noise and phase delay among the vectors as was proposed by (Martins et al., 2006).

LOAD TORQUE ESTIMATION - KALMAN FILTER

The reduced mechanical IM system can be represented by the following equations

$$\frac{d}{dt}\begin{bmatrix}\omega \\ T_L\end{bmatrix} = \begin{bmatrix}-\frac{B_n}{J} & -\frac{1}{J} \\ 0 & 0\end{bmatrix}\begin{bmatrix}\omega \\ T_L\end{bmatrix} + \begin{bmatrix}\frac{1}{J} \\ 0\end{bmatrix}T_e \tag{33}$$

$$y = \begin{bmatrix}1 & 0\end{bmatrix}\begin{bmatrix}\omega \\ T_L\end{bmatrix} \tag{34}$$

The Kalman Filter could be used to provide the value of torque load or disturbances - T_L. Since (15)-(16) is nonlinear, the Kalman filter linearizes the model at the actual operating point (Aström & Wittenmark, 1997). In addition, this filter takes into account the signal noise, which could be generated as pulse width modulation drivers. Assuming the definitions

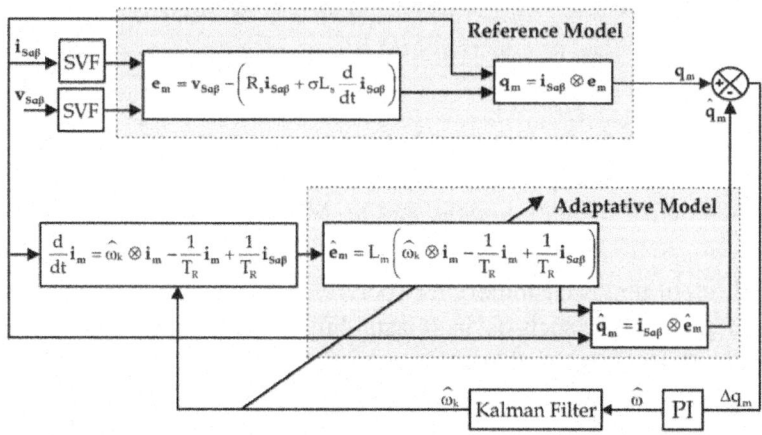

Figure 3: Reactive Power MRAS Speed Estimation

$$x_k = \begin{bmatrix} \hat{\omega}_k & \hat{T}_L \end{bmatrix}^T, \mathbf{A_m} = \begin{bmatrix} -\dfrac{B_n}{J} & -\dfrac{1}{J} \\ 0 & 0 \end{bmatrix}, \mathbf{B_m} = \begin{bmatrix} \dfrac{1}{J} \\ 0 \end{bmatrix}, \mathbf{C_m} = \begin{bmatrix} 1 & 0 \end{bmatrix} \text{ and } y_k = \hat{\omega}.$$

Then, the recursive equation for the discrete time Kalman Filter (De Campos et al., 2000) is described by

$$\mathbf{K}(k) = \mathbf{P}(k)\mathbf{C_m^T}\left(\mathbf{C_m}\mathbf{P}(k)\mathbf{C_m^T} + \mathbf{R}\right)^{-1} \tag{35}$$

where K(k) is the Kalman gain. The covariance matrix P(k) is given by

$$\mathbf{P}(k+1) = (\mathbf{I} - \mathbf{A_m}t_s)\left(\mathbf{P}(k) - \mathbf{K}(k)\mathbf{C_m}\mathbf{P}(k)\right)(\mathbf{I} - \mathbf{A_m}t_s)^T + (\mathbf{B_m}t_s)\mathbf{Q}(\mathbf{B_m}t_s)^T \tag{36}$$

Therefore, the estimated torque $\hat{T}_L$ is one observed state of the Kalman filter

$$\hat{\mathbf{x}}_{\mathbf{k}}(k+1) = (\mathbf{I} - \mathbf{A_m}t_s)\hat{\mathbf{x}}_{\mathbf{k}}(k) + \mathbf{B_m}t_s u(k) + (\mathbf{I} - \mathbf{A_m}t_s)\mathbf{K}(k)\left(\hat{\omega} - \mathbf{C_m}\hat{\mathbf{x}}_{\mathbf{k}}(k)\right) \tag{37}$$

giving $\hat{\omega} \approx \hat{\omega}_k$ and $\hat{\mathbf{x}}_{\mathbf{k}}(k) = \begin{bmatrix} \hat{\omega}_k & T_L \end{bmatrix}^T$.

The matrices R and Q are defined according to noise elements of predicted state variables, taking into account the measurement noise covariance R and the plant noise covariance Q.

STATE VARIABLE FILTER

The state variable filter (SVF) is used to mathematically evaluate differentiation signals. This filter is necessary in the implementation of FLC and MRAS algorithms. The transfer function of SVF is of second order as it is necessary to obtain the first order derivative.

$$G_{svf} = \frac{\omega_{svf}}{(s+\omega_{svf})^2} \qquad (38)$$

where $\omega_{sv\,f}$ is the filter bandwidth defined at around 5 to 10 times the input frequency signal $u_{sv\,f}$.

The discretized transfer function, using the Euler method, can be performed in state-space as

$$\mathbf{x_{svf}}(k+1) = \mathbf{A_{svf}}\mathbf{x_{svf}}(k) + \mathbf{B_{svf}}u_{svf}(k) \qquad (39)$$

where

$$\mathbf{A_{svf}} = \begin{bmatrix} 1 & 1 \\ -\omega_{svf}^2 & 1-2\omega_{svf} \end{bmatrix}, \mathbf{B_{svf}} = \begin{bmatrix} 0 \\ \omega_{svf}^2 \end{bmatrix} \text{ and } \mathbf{x_{svf}} = \begin{bmatrix} x_1 \\ x_2 \end{bmatrix}.$$

The state variables x_1 and x_2 represent the input filtered signal and input derivative signal.

EXPERIMENTAL RESULTS

Sensorless control schemes were implemented in DSP based platform using TMS 320F2812. Experimental results were carried out on a motor with specifications: 1.5cv, 380V, 2.56A, 60 Hz, R_s = 3.24Ω, R_r = 4.96Ω, L_r = 404.8mH, L_s = 402.4mH, L_m = 388.5mH, N = 2 and nominal speed of 188 rad/s.

The experimental analyses are carried out with the following operational sequence:
- The motor is excited (during 10 s to 12 s) using a smooth flux reference trajectory.
- Starting from zero initial value, the rotor speed reference grows linearly until it reaches the reference value. Thus, the reference rotor speed value is kept constant.
- During stand-state, a step constant load torque is applied.

In order to generate load variation for torque disturbance analyses, the DC motor is connected to an induction motor driving-shaft. Then, the load shaft varies in accordance with DC motor field voltage and inserting a resistance on its armature. Fig. 4 and Fig. 5 depict performance of both control schemes:

FLC control and simplified FLC control with rotor speed reference of 18 rad/s. In these figures measure speed, estimated speed, stator (q-d) currents and estimated torque are illustrated.

Fig.6 and Fig. 7 present experimental results with 36 rad/s rotor speed reference. Fig. 8 and Fig. 9 show FLC control and Simplified FLC with 45 rad/s rotor speed reference. The above figures present experimental results for low rotor speed range of FLC control and Simplified FLC control applying load torque. In accordance with the figures above, both control schemes present similar performances in steady state. It is verified that both schemes respond to compensated torque variations. With respect to Simplified FLC, it is necessary to carefully select fixed gains in order to guarantee the alignment of the rotor flux on the d axis.

CONCLUSION

Two different sensorless IM control schemes were proposed and developed based on nonlinear control - FLC Control and Simplified FLC Control. These control schemes are composed of a flux-speed controller, which is derived from a fifth-order IM model. In the implementation of feedback linearization control (FLC), the control algorithm presents a large number of computational requirements. In the simplified FLC scheme, a substitution of FLC currents controllers by two PI controllers is proposed to generate the stator drive voltage. In order to provide the rotor speed for both control schemes, a MRAS algorithm based on reactive power is applied.

To correctly evaluate whether this Simplified FLC does not affect control performance, a comparative experimental analysis of a FLC control and a simplified FLC control is presented. Experimental results in DSP TMS 320F2812 platform show the performance of both systems in the 18rad/s, 36 rad/s and 45 rad/s rotor speed range. Both control schemes present similar performance in steady-state. Hence, the proposed modification of the FLC control allows a simplification of the control algorithm without deterioration in control performance. However, it may necessary to carefully evaluate the gain selection in the simplified FLC control, to guarantee rotor flux alignment on the d axis, as well as, to guarantee speed-flux decoupling. Both control schemes indicate sensitivity with model parameter variation, and one way to overcome this would be the is development of an adaptive FLC control laws on FLC control.

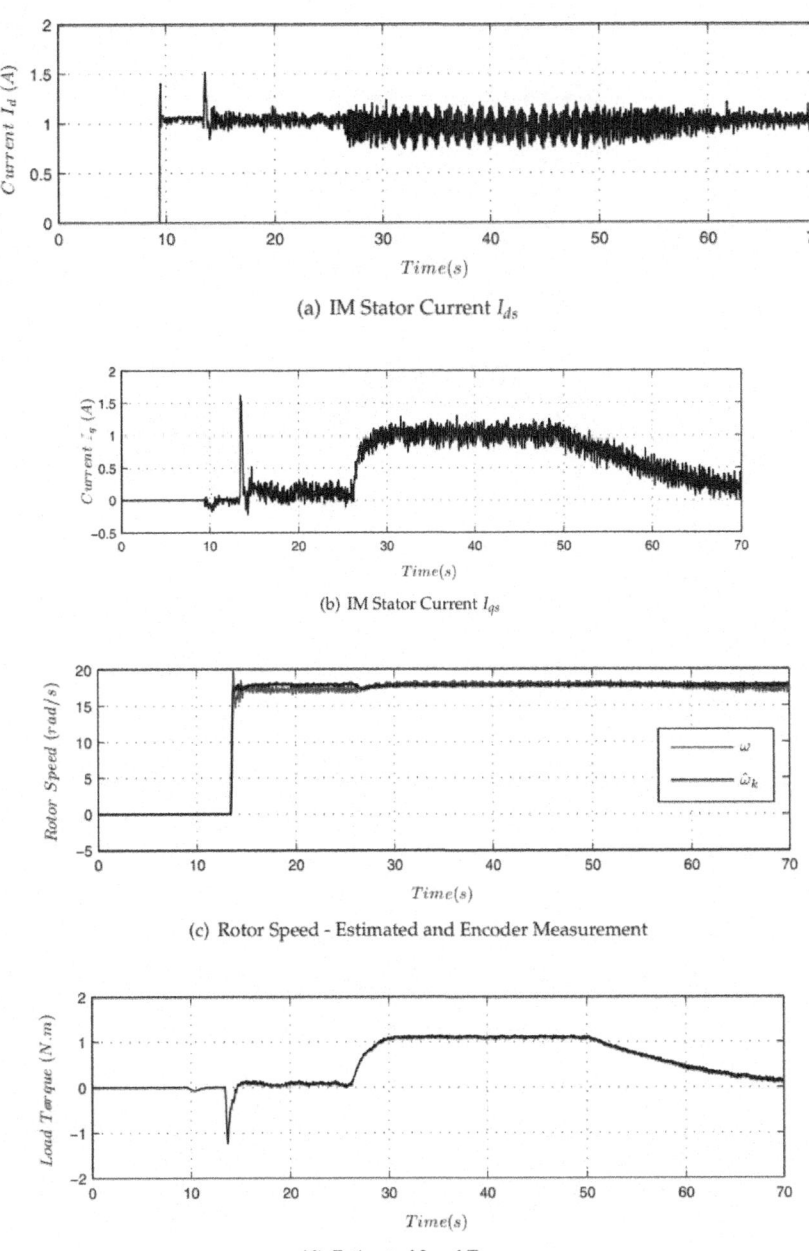

(a) IM Stator Current I_{ds}

(b) IM Stator Current I_{qs}

(c) Rotor Speed - Estimated and Encoder Measurement

(d) Estimated Load Torque

Figure 4: FLC control with 18 rad/s rotor speed reference

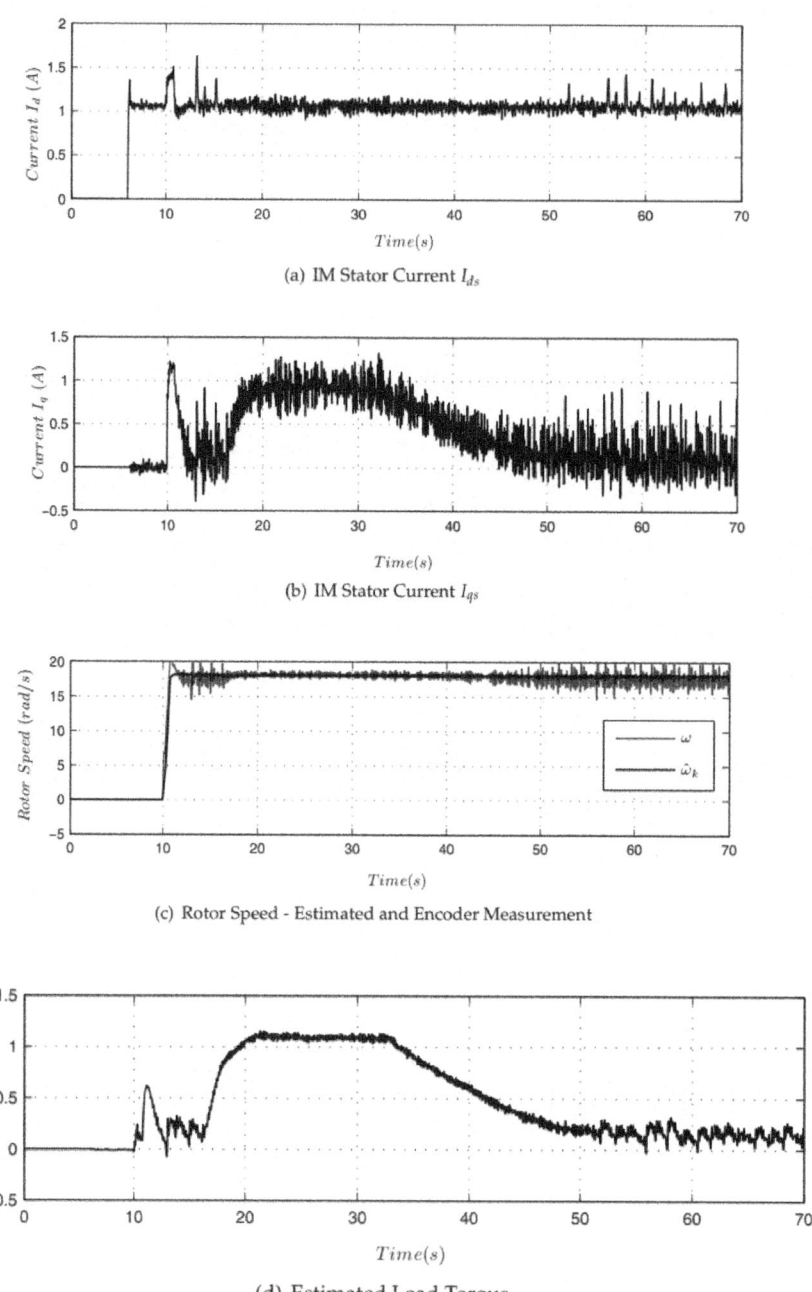

(a) IM Stator Current I_{ds}

(b) IM Stator Current I_{qs}

(c) Rotor Speed - Estimated and Encoder Measurement

(d) Estimated Load Torque

Figure 5: Simplified FLC control with 18 rad/s rotor speed reference

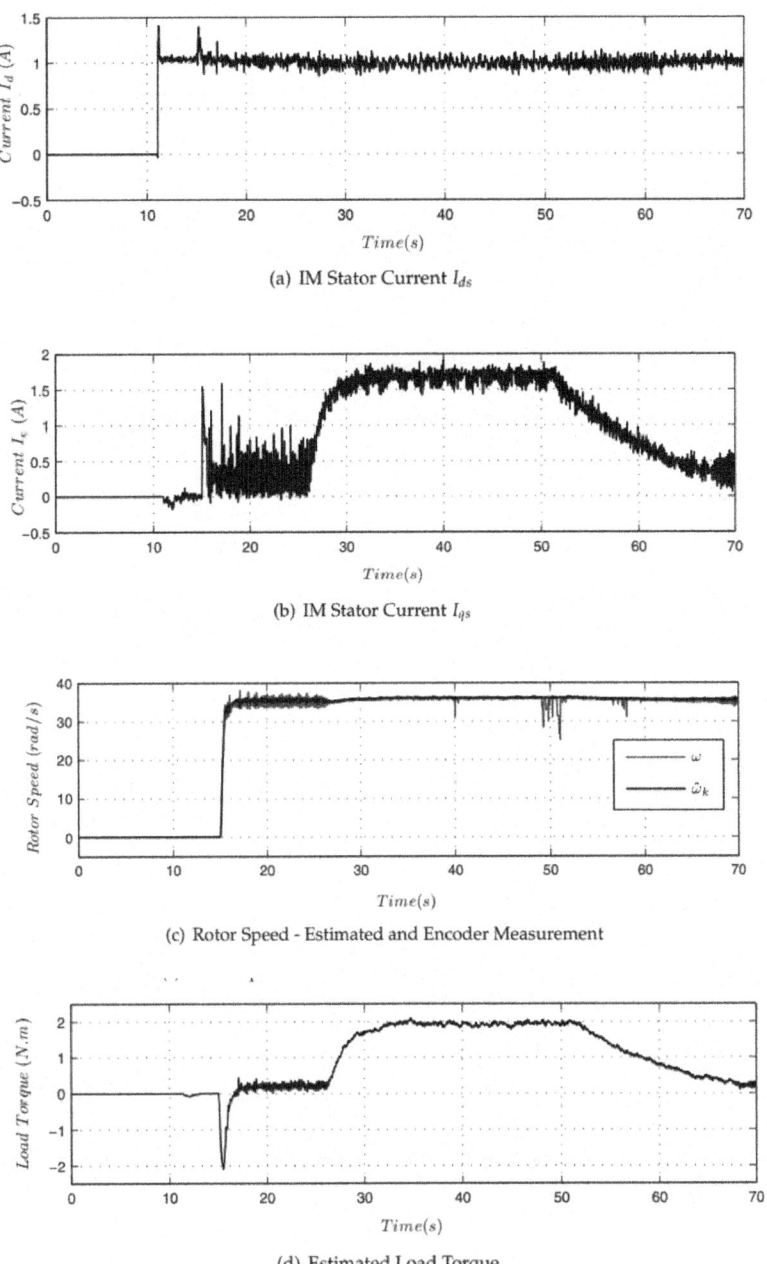

(a) IM Stator Current I_{ds}

(b) IM Stator Current I_{qs}

(c) Rotor Speed - Estimated and Encoder Measurement

(d) Estimated Load Torque

Figure 6: FLC control with 36 rad/s rotor speed reference

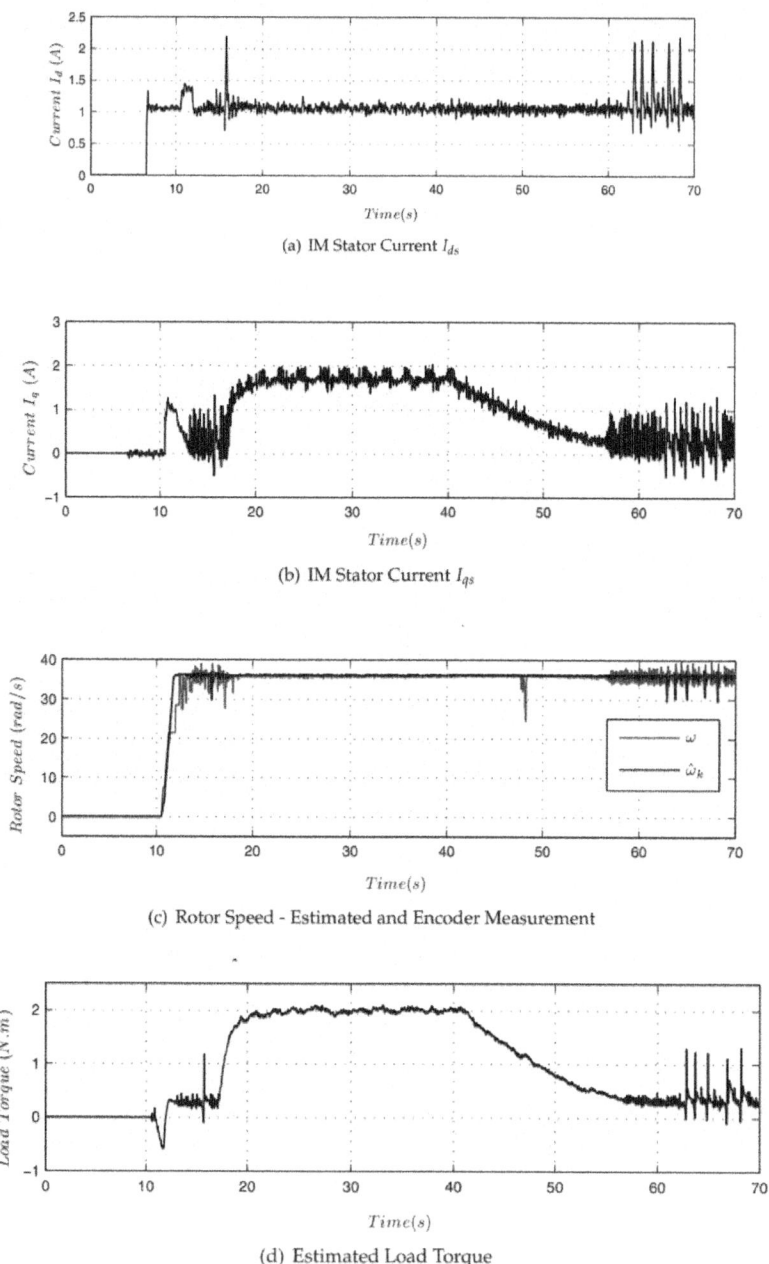

(a) IM Stator Current I_{ds}

(b) IM Stator Current I_{qs}

(c) Rotor Speed - Estimated and Encoder Measurement

(d) Estimated Load Torque

Figure 7: Simplified FLC control with 36 rad/s rotor speed reference

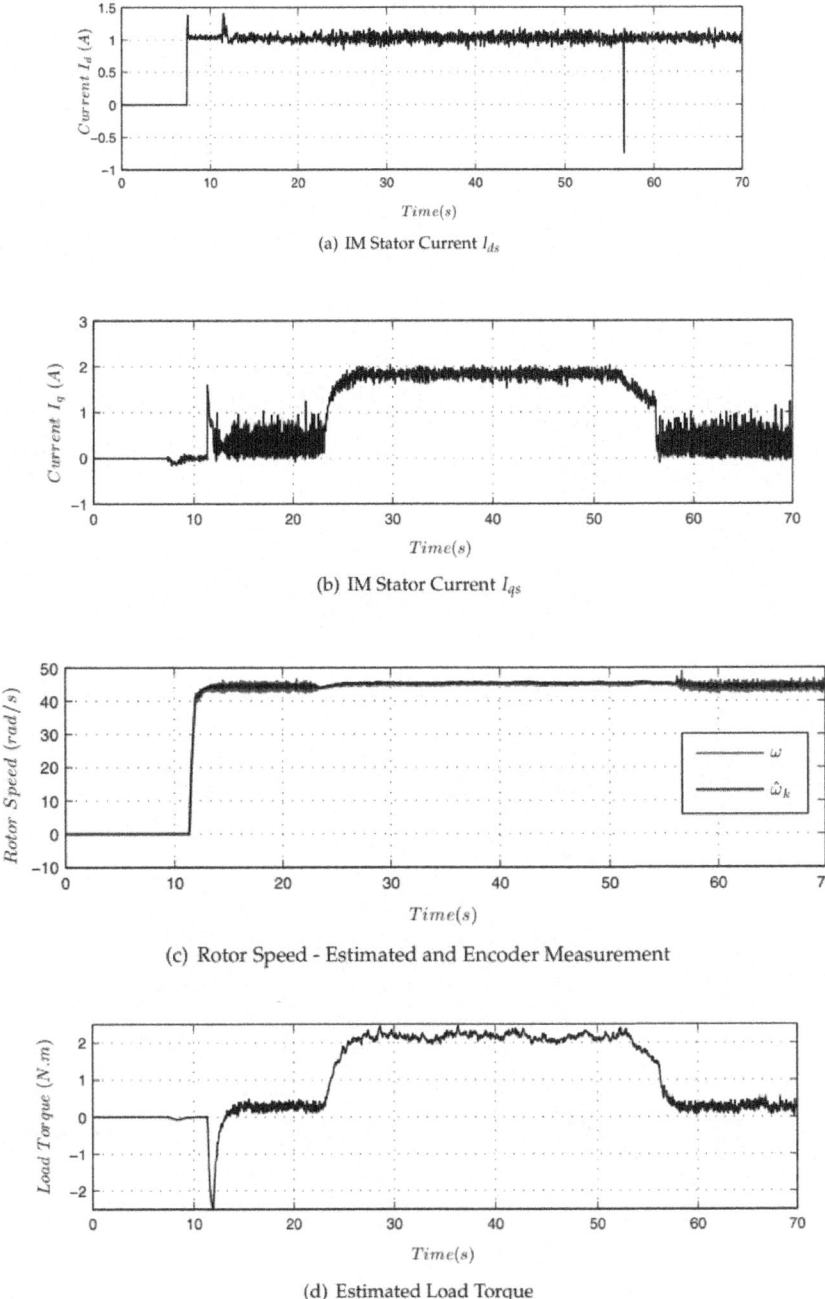

(a) IM Stator Current I_{ds}

(b) IM Stator Current I_{qs}

(c) Rotor Speed - Estimated and Encoder Measurement

(d) Estimated Load Torque

Figure 8: FLC control with 45 rad/s rotor speed reference

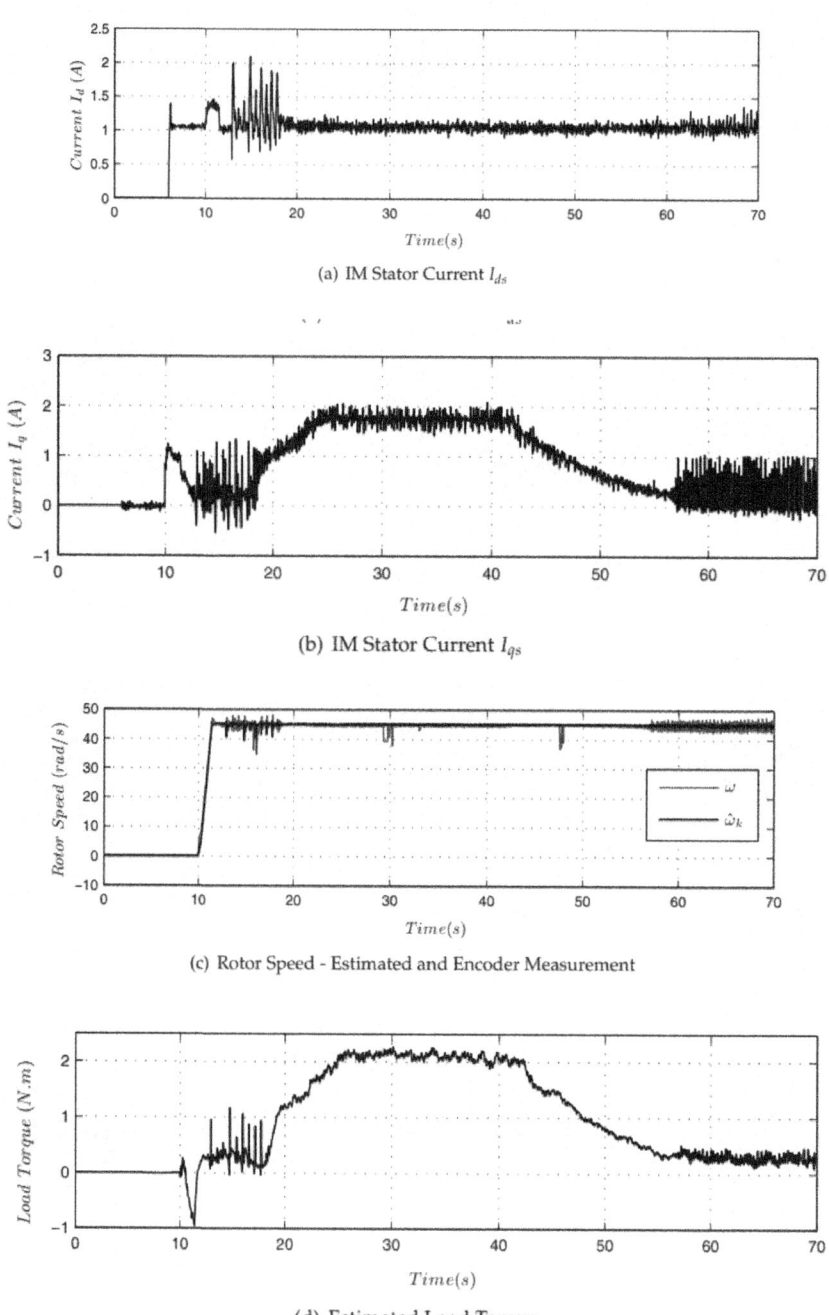

(a) IM Stator Current I_{ds}

(b) IM Stator Current I_{qs}

(c) Rotor Speed - Estimated and Encoder Measurement

(d) Estimated Load Torque

Figure 9: Simplified FLC control with 45 rad/s rotor speed reference

REFERENCES

1. Aström, K. & Wittenmark, B. (1997). Computer-Controlled Systems: Theory and Design, Prentice-Hall.
2. Cardoso, R. & Gründling, H. A. (2009). Grid synchronization and voltage analysis based on the kalman filter, in V. M. Moreno & A. Pigazo (eds), Kalman Filter Recent Advances and Applications, InTech, Croatia, pp. 439–460.
3. De Campos, M., Caratti, E. & Grundling, H. (2000). Design of a position servo with induction motor using self-tuning regulator and kalman filter, Conference Record of the 2000 IEEE Industry Applications Conference, 2000.
4. Gastaldini, C. & Grundling, H. (2009). Speed-sensorless induction motor control with torque compensation, 13th European Conference on Power Electronics and Applications, EPE '09, pp. 1–8.
5. Krause, P. C. (1986). Analysis of electric machinery, McGraw-Hill.
6. Leonhard, W. (1996). Control of Electrical Drives, Springer-Verlag.
7. Marino, R., Peresada, S. & Valigi, P. (1990). Adaptive partial feedback linearization of induction motors, Proceedings of the 29th IEEE Conference on Decision and Control, 1990, pp. 3313–3318 vol.6.
8. Marino, R., Tomei, P. & Verrelli, C. M. (2004). A global tracking control for speed-sensorless induction motors, Automatica 40(6): 1071–1077.
9. Martins, O., Camara, H. & Grundling, H. (2006). Comparison between mrls and mras applied to a speed sensorless induction motor drive, 37th IEEE Power Electronics Specialists Conference, PESC '06., pp. 1–6.
10. Montanari, M., Peresada, S., Rossi, C. & Tilli, A. (2007). Speed sensorless control of induction motors based on a reduced-order adaptive observer, IEEE Transactions on Control Systems Technology 15(6): 1049–1064.
11. Montanari, M., Peresada, S. & Tilli, A. (2006). A speed-sensorless indirect field-oriented control for induction motors based on high gain speed estimation, Automatica 42(10): 1637–1650.
12. Orlowska-Kowalska, T. & Dybkowski, M. (2010). Stator-current-based mras estimator for a wide range speed-sensorless induction-motor drive, IEEE Transactions on Industrial Electronics 57(4): 1296–1308.
13. Peng, F.-Z. & Fukao, T. (1994). Robust speed identification for speed-sensorless vector control of induction motors, IEEE Transactions on Industry Applications 30(5): 1234–1240.
14. Peresada, S. & Tonielli, A. (2000). High-performance robust speed-flux

tracking controller for induction motor, International Journal of Adaptive Control and Signal Processing, 2000.
15. Vieira, R., Azzolin, R. & Grundling, H. (2009). A sensorless single-phase induction motor drive with a mrac controller, 35th Annual Conference of IEEE Industrial Electronics, IECON '09., pp. 1003–1008.

Chapter 9

A RMRAC PARAMETER IDENTIFICATION ALGORITHM APPLIED TO INDUCTION MACHINES

Rodrigo Z. Azzolin[1], Cristiane C. Gastaldini[2], Rodrigo P. Vieira[3], and Hilton A. Gründling[4]

[1,2,3,4]Federal University of Santa Maria
[1]Federal University of Rio Grande
[3]Federal University of Pampa Brazil

INTRODUCTION

This chapter deals with the problem of parameter identification of electrical machines to achieve good performance of a control system. The development of an identification algorithm is presented, which in this case is applied to Single-Phase Induction Motors, but could easily be applied to other electrical machine. This scheme is based on a Robust Model Reference Adaptive Controller and measurements of stator currents of a machine with standstill rotor. From the obtained parameters, it is possible to design high performance controllers and sensorless control for induction motors.

Single-phase induction motors (SPIM) are widely used in fractional and sub-fractional horsepower applications, usually in locations where only single-phase energy supply is available. In most of these applications the machine operates at constant frequency and is fed directly from the AC grid with an ON/OFF starting procedure. In recent years, several researchers have shown that the variable speed operation can enhance the SPIM's efficiency (Blaabjerg et al., 2004; Donlon et al., 2002). In addition, other researchers have developed high performance drives for SPIM's, using Field-Oriented Control (FOC) and sensorless techniques (de Rossiter Correa et al., 2000; Vaez-Zadeh & Reicy, 2005; Vieira et al., 2009b). However, the FOC associated with the sensorless technique demands accurate knowledge of electrical motor parameters to achieve good performance.

A good deal of research has been carried out in the last several years on parameter estimation of induction motors, mainly with regards to three-phase induction machines (Azzolin et al., 2007; Koubaa, 2004; Ribeiro et al., 1995; Velez-Reyes et al., 1989).

However, few methods have been proposed for automatic estimation in single-phase induction motors. One of them uses a classical method for electrical parameter identification (Ojo & Omozusi, 2001), its implementation is onerous. In (Vieira et al., 2009a) a Recursive Least Squares (RLS) identification algorithm is used to obtain machine parameters. The identification results are good, although this method requires the design of filters to obtain the variable derivative, which can be deteriorated by noises.

In order to solve these problems, this chapter details a closed-loop algorithm to estimate the electrical parameters of a single-phase induction motor based on (Azzolin & Gründling, 2009). In (Azzolin & Gründling, 2009) a Robust Model Reference Adaptive Controller (RMRAC) algorithm was used for parameter identification of three-phase induction motors. The RMRAC algorithm eliminates the use of filters to obtain derivatives of the signals. Thus, the objectives of this chapter are to apply the RMRAC algorithm in the electrical parameter identification of a SPIM and make use of the robustness of the system in dealing with of noise measurement.

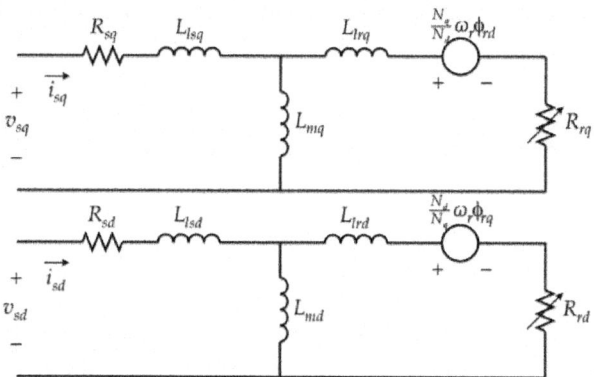

Figure 1: Equivalent circuit of a SPIM.

As a result, the proposed parameter estimation procedure is divided into three steps:
- identification of the RMRAC controller gains;
- estimation of the transfer function coefficients of the induction motor at standstill;

- calculation of the stator resistance R_{si}, rotor resistance R_{ri}, stator inductance L_{si}, rotor inductance L_{ri} and mutual inductance L_{mi} using steps (I) and (II), where the index "$_i$" express the axes q or d.

This chapter is organized as follows: Section 2 presents the induction machine model. A short review of the RMRAC algorithm applied to the identification system is presented in sections 3 and 4. Section 5 describes the assumptions and equations of Model Reference Control (MRC) while section 6 shows the proposed parameter identification algorithm. Sections 7 and 8 present the simulation and experimental results. Finally, chapter conclusions are presented in section 9.

MATHEMATICAL MODEL OF A SINGLE PHASE INDUCTION MOTOR

The SPIM equivalent circuit without the permanent split-capacitor can be represented by an asymmetrical two-phase induction motor as shown in Fig. 1. In this figure L_{lsi} and L_{lri} are the stator and rotor leakage inductances, ω_r is the speed rotor, φ_{ri} is the electromagnetic flux and N_i is the number of turns for auxiliary winding or axis d and for main winding or axis q. The stator and rotor inductances are relationship with the leakage and mutual inductance as

a $L_{si} = L_{lsi} + L_{mi}$ and $L_{ri} = L_{lri} + L_{mi}$, respectively.

From Fig. 1 and from (Krause et al., 1986) it is possible to derive the dynamical model of a SPIM. The SPIM dynamical model in a stationary reference frame can be described by 1. In this equation $\bar{\sigma}_q = L_{sq}L_{rq} - L_{mq}^2$, $\bar{\sigma}_d = L_{sd}L_{rd} - L_{md}^2$, p is the poles pairs and n is the relationship between the number of turns for auxiliary and for main winding N_d/N_q.

$$\begin{bmatrix} \dot{i}_{sq} \\ \dot{i}_{sd} \\ \dot{i}_{rq} \\ \dot{i}_{rd} \end{bmatrix} = \begin{bmatrix} -\dfrac{R_{sq}L_{rq}}{\bar{\sigma}_q} & -p\omega_r \dfrac{1}{n}\dfrac{L_{mq}L_{md}}{\bar{\sigma}_q} & \dfrac{R_{rq}L_{mq}}{\bar{\sigma}_q} & -p\omega_r \dfrac{1}{n}\dfrac{L_{rd}L_{mq}}{\bar{\sigma}_q} \\ p\omega_r n\dfrac{L_{mq}L_{md}}{\bar{\sigma}_d} & -\dfrac{L_{rd}R_{sd}}{\bar{\sigma}_d} & p\omega_r n\dfrac{L_{rq}L_{md}}{\bar{\sigma}_d} & \dfrac{R_{rd}L_{md}}{\bar{\sigma}_d} \\ \dfrac{L_{mq}R_{sq}}{\bar{\sigma}_q} & p\omega_r \dfrac{1}{n}\dfrac{L_{sq}L_{md}}{\bar{\sigma}_q} & -\dfrac{L_{sq}R_{rq}}{\bar{\sigma}_q} & p\omega_r \dfrac{1}{n}\dfrac{L_{sq}L_{rd}}{\bar{\sigma}_q} \\ -p\omega_r n\dfrac{L_{sd}L_{mq}}{\bar{\sigma}_d} & \dfrac{L_{md}R_{sd}}{\bar{\sigma}_d} & -p\omega_r n\dfrac{L_{sd}L_{rq}}{\bar{\sigma}_d} & -\dfrac{L_{sd}R_{rd}}{\bar{\sigma}_d} \end{bmatrix} \cdot \begin{bmatrix} i_{sq} \\ i_{sd} \\ i_{rq} \\ i_{rd} \end{bmatrix} + \begin{bmatrix} \dfrac{L_{rq}}{\bar{\sigma}_q} & 0 \\ 0 & \dfrac{L_{rd}}{\bar{\sigma}_d} \\ -\dfrac{L_{mq}}{\bar{\sigma}_q} & 0 \\ 0 & -\dfrac{L_{md}}{\bar{\sigma}_d} \end{bmatrix} \begin{bmatrix} v_{sq} \\ v_{sd} \end{bmatrix}$$

(1)

From equation 1 it is possible to obtain the transfer functions in the axes q and d at standstill rotor ($\omega_r = 0$), where these equations are decoupled and presented in 2 and 3.

$$H_q(s) = \frac{i_{sq}(s)}{v_{sq}(s)} = \frac{s\frac{L_{rq}}{\bar{\sigma}_q} + \frac{R_{rq}}{\bar{\sigma}_q}}{s^2 + sp_q + \frac{R_{rq}R_{sq}}{\bar{\sigma}_q}}, \tag{2}$$

$$H_d(s) = \frac{i_{sd}(s)}{v_{sd}(s)} = \frac{s\frac{L_{rd}}{\bar{\sigma}_d} + \frac{R_{rd}}{\bar{\sigma}_d}}{s^2 + sp_d + \frac{R_{rd}R_{sd}}{\bar{\sigma}_d}}, \tag{3}$$

where

$$p_q = \frac{R_{sq}L_{rq} + R_{rq}L_{sq}}{\bar{\sigma}_q} \text{ and } p_d = \frac{R_{sd}L_{rd} + R_{rd}L_{sd}}{\bar{\sigma}_d}. \tag{4}$$

IDENTIFICATION SYSTEM

The proposed electrical parameter identification system is shown in Fig. 2. This system is based on a RMRAC algorithm used to generate the control action v_{sq} by the difference between the measured current i_{sq} and the reference i^*_{sq} at standstill rotor. In this test the auxiliary winding (or axis d) is open while the main winding (or axis q) is identified.

The dotted box in Fig. 2 is detailed in Fig. 3 where the RMRAC control law applied to q axis is shown. In Fig. 3, the reference model and plant are given by

$$W_m(s) = k_m \frac{Z_m(s)}{R_m(s)}, \tag{5}$$

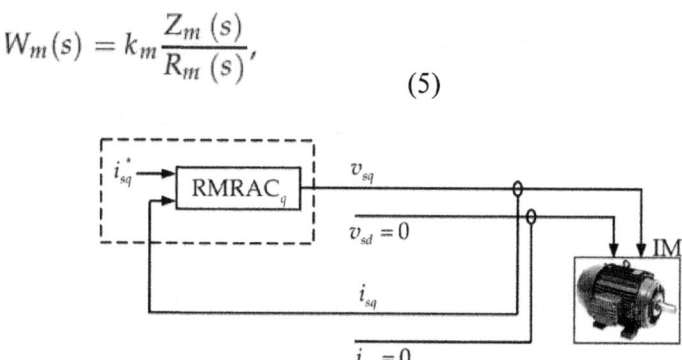

Figure 2: Block diagram of the RMRAC identification system.

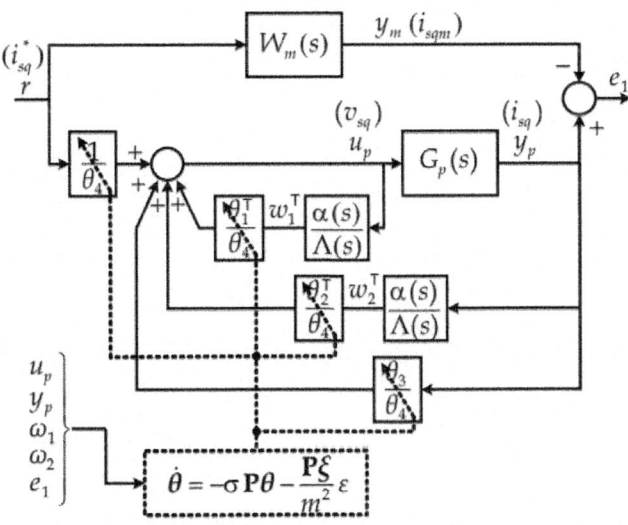

Figure 3: RMRAC structure

$$G_p(s) = k_p \frac{Z_p(s)}{R_p(s)}. \qquad (6)$$

where $Z_m(s)$, $R_m(s)$, $Z_p(s)$ and $R_p(s)$ are monic polynomials and k_m and k_p are high frequency gains.

The general idea behind RMRAC is to create a closed-loop controller with gains that can be updated and change the response of the system. The output of the system y_p is compared to a desired response from a reference model y_m. The controller gains vector $\theta = [\theta_1^T \ \theta_2^T \ \theta_3 \ \theta_4]^T$ is updated by a gradient algorithm based on e_1 error. The goal is that the gains converge to ideal values that cause the plant response to match the response of the reference model. These gains can be obtained by a gradient algorithm presented in (Ioannou & Sun, 1996; Ioannou & Tsakalis, 1986) and described as follows. More details of RMRAC algorithms can be seen in (Câmara & Gründling, 2004).

The procedure described in Fig. 2 and Fig. 3 is applied to parameter identification of the q axis or transfer function 2. However, the same procedure is used for parameter identification of the d axis or transfer function 3.

RMRAC GAINS ADAPTATION ALGORITHM

The gradient algorithm used to obtain the control law gains is given by

$$\dot{\theta} = -\sigma P\theta - \frac{P\zeta}{m^2}\varepsilon, \tag{7}$$

with

$$\dot{m} = \delta_0 m + \delta_1 (|u_p| + |y_p| + 1), \quad m(0) > \frac{\delta_1}{\delta_0}, \quad \delta_1 \geq 1, \tag{8}$$

and

$$\zeta = W_m(s)Iw, \tag{9}$$

$$w = [w_1^T \; w_2^T \; y_p \; u_p]^T, \tag{10}$$

w_1^T, w_2^T are auxiliary vectors, δ_0, δ_1 are positive constants and δ_0 satisfies $\delta_0 + \delta_2 \leq \min(p_0, q_0)$, $q_0 \in \Re^+$ is such that the $W_m(s - q_0)$ poles and the $(F - q0I)$ eigenvalues are stable and δ_2 is a positive constant. The sigma modification σ in 7 is given by

$$\sigma = \begin{cases} 0 & \text{if} \quad \|\theta\| < M_0 \\ \sigma_0 \left(\frac{\|\theta\|}{M_0} - 1 \right) & \text{if} \quad M_0 \leq \|\theta\| < 2M_0 , \\ \sigma_0 & \text{if} \quad \|\theta\| \geq 2M_0 \end{cases} \tag{11}$$

where $M_0 > \|\theta^*\|$ and $\sigma_0 > 2\mu^{-2}/R^2, R, \mu \in \Re^+$ are design parameters. In this case, the parameters used in the implementation of the gradient algorithm are

$$\begin{cases} \delta_0 = 0.7 \\ \delta_1 = 1 \\ \delta_2 = 1 \\ \sigma_0 = 0.1 \\ M_0 = 10 \end{cases} \tag{12}$$

More details of design of the gradient algorithm can be seen in (Ioannou & Tsakalis, 1986). As defined in (Lozano-Leal et al., 1990), the modified error in 7 is given by

$$\varepsilon = e_1 + \theta^T \zeta - W_m \theta^T w, \tag{13}$$

or

$$\varepsilon = \phi^T \zeta + \mu\eta. \tag{14}$$

When the ideal values of gains are identified and the plant model is well known, the plant can be obtained by equation analysis of MRC algorithm described in the next section.

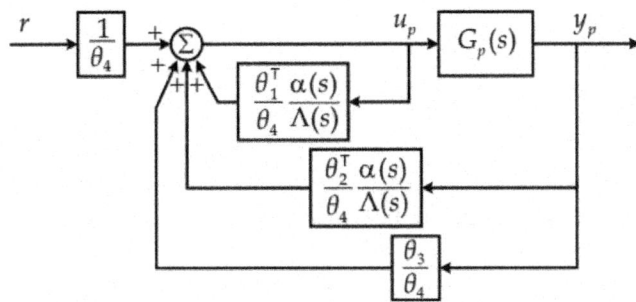

Figure 4: MRC structure.

MRC ANALYSIS

The Model Reference Control (MRC) shown in Fig. 4 can be understood as a particular case of RMRAC structure, which is presented in Fig. 3. This occurs after the convergence of the controller gains when the gradient algorithm changes to the steady-state. It is important to note that this analysis is only valid when the plant model is well known and free of unmodeled dynamics and parametric variations.

To allow the analysis of the MRC structure, the plant and reference model must satisfy some assumptions as verified in (Ioannou & Sun, 1996). These suppositions, which are also valid for RMRAC, are given as follow:

Plant Assumptions:

P1. $Z_p(s)$ is a monic Hurwitz polynomial of degree m_p;

P2. An upper bound n of the degree n_p of $R_p(s)$;

P3. The relative degree $n^* = n_p - m_p$ of $G_p(s)$;

P4. The signal of the high frequency gain k_p is known.

Reference Model Assumptions:

M1. $Z_m(s)$, $R_m(s)$ are monic Hurwitz polynomials of degree q_m, p_m, respectively, where $p_m \leq n$;

M2. The relative degree $n_m^* = p_m - q_m$ of $W_m(s)$ is the same as that of $G_p(s)$, i.e, $n_m^* = n^*$.

In Fig. 4 the feedback control law is

$$u_p = \frac{\theta_1^T \alpha(s)}{\theta_4 \Lambda(s)} u_p + \frac{\theta_2^T \alpha(s)}{\theta_4 \Lambda(s)} y_p + \frac{\theta_3}{\theta_4} y_p + \frac{1}{\theta_4} r, \tag{15}$$

and

$$\alpha(s) \triangleq \alpha_{n-2}(s) = \left[s^{n-2}, s^{n-3}, \ldots, s, 1\right]^T \quad \text{for} \quad n \geq 2,$$
$$\alpha(s) \triangleq 0 \quad \text{for} \quad n = 1, \tag{16}$$

$\theta_3, \theta_4 \in \Re^1$; $\theta_1^T, \theta_2^T \in \Re^{n-1}$ are constant parameters to be designed and $\Lambda(s)$ is an arbitrary monic Hurwitz polynomial of degree $n-1$ that contains $Z_m(s)$ as a factor, i.e.,

$$\Lambda(s) = \Lambda_0(s) Z_m(s), \tag{17}$$

which implies that $\Lambda_0(s)$ is monic, Hurwitz and of degree $n_0 = n - 1 - q_m$. The controller parameter vector is given so that the closed loop plant from r to y_p is equal to $W_m(s)$.

$$\theta = \left[\theta_1^T \; \theta_2^T \; \theta_3 \; \theta_4\right]^T \in \Re^{2n}, \tag{18}$$

The I/O properties of the closed-loop plant shown in Fig. 4 are described by the transfer function equation

$$y_p = G_c(s) r, \tag{19}$$

where

$$G_c(s) = \frac{k_p Z_p \Lambda^2}{\Lambda \left[(\theta_4 \Lambda - \theta_1^T \alpha) R_p - k_p Z_p (\theta_2^T \alpha + \theta_3 \Lambda)\right]}, \tag{20}$$

Now, the objective is to choose the controller gains so that the poles are stable and the closed-loop transfer function $G_c(s) = W_m(s)$, i.e.,

$$\frac{k_p Z_p \Lambda^2}{\Lambda \left[(\theta_4 \Lambda - \theta_1^T \alpha) R_p - k_p Z_p (\theta_2^T \alpha + \theta_3 \Lambda)\right]} = k_m \frac{Z_m}{R_m}. \tag{21}$$

Thus, considering a system free of unmodeled dynamics, the plant coefficients can be known by the MRC structure, i.e., k_p, $Z_p(s)$ and $R_p(s)$ are given by 21 when the controller gains θ_1^T, θ_2^T, θ_3 and θ_4 are known and $W_m(s)$ is previously defined.

PARAMETER IDENTIFICATION USING RMRAC

The proposed parameter estimation method is executed in three steps, described as follows:

First step: Convergence of controller gains vector

The proposed parameter identification method is shown in Fig. 2. In this figure the parameter identification of q axis is shown, but the same procedure is performed for parameter identification of d axis, one procedure at a time.

A Persistent Excitant (PE) reference current i_{sq}^* is applied at q axis of SPIM at standstill rotor. The current i_{sq} is measured and controlled by the RMRAC structure while i_{sd} stays at null value. The controller structure is detailed in Fig. 3. When e_1 goes to zero, the controller gains go to an ideal value. Subsequently, the gradient algorithm is put in steady-state and the system looks like the MRC structure given by Fig. 4. Therefore, the transfer function coefficients can be found using equation 21.

Second step: Estimation of k_{pi}, h_{0i}, a_{1i} and a_{0i}

This step consists of the determination of the Linear-Time-Invariant (LTI) model of the induction motor. The machine is at standstill and the transfer functions given in 2 and 3 can be generalized as follows

$$\frac{i_{si}}{v_{si}} = k_{pi} \frac{Z_{pi}(s)}{R_{pi}(s)} = k_{pi} \frac{s + h_{0i}}{s^2 + sa_{1i} + a_{0i}}, \tag{22}$$

where

$$k_{pi} = \frac{L_{ri}}{\bar{\sigma}_i}, \quad h_{0i} = \frac{R_{ri}}{L_{ri}}, \quad a_{1i} = p_i \text{ and } a_{0i} = \frac{R_{si} R_{ri}}{\bar{\sigma}_i}, \tag{23}$$

The reference model given by 5 is rewritten as

$$W_m(s) = k_m \frac{Z_m}{R_m} = k_m \frac{s + z_0}{s^2 + p_1 s + p_0}, \tag{24}$$

and from the plant and reference model assumptions results

$$\begin{cases} m_p = 1, \, n_p = 2, \, n^* = 1, \\ q_m - 1, \, p_m - 2, \, n_m^* - 1, \end{cases} \tag{25}$$

The upper bound n is chosen equal to n_p because the plant model is considered well known and with $n = n_p$ only one solution is guaranteed for the controller gains. Thus, the filters are given by

$$\begin{cases} \Lambda(s) = Z_m(s) = s + z_0, \\ \alpha(s) = z_0, \end{cases} \tag{26}$$

Assuming the complete convergence of controller gains, the plant coefficients are obtained combining the equations 22, 24 and 26 in 21 and are given by

$$\begin{cases} k_{pi} = k_m \theta_{4i}, \\ h_{0i} = \frac{z_0}{\theta_{4i}} \left(\theta_{4i} - \theta_{1i}^T \right), \\ a_{1i} = p_1 + k_m \theta_{3i}, \\ a_{0i} = p_0 + k_m z_0 \left(\theta_{2i}^T + \theta_{3i} \right). \end{cases} \quad (27)$$

Third step: R_{si}, R_{ri}, L_{si}, L_{ri} and L_{mi} calculation

Combining the equations 4, 23 and using the values obtained in 27 after the convergence of the controller gains, we obtain the parameters of the induction motor:

$$\begin{cases} \hat{R}_{si} = \frac{a_{0i}}{k_{pi} h_{0i}}, \\ \hat{R}_{ri} = \frac{a_{1i}}{k_{pi}} - \hat{R}_{si}, \\ \hat{L}_{si} = \hat{L}_{ri} = \frac{\hat{R}_{ri}}{h_{0i}}, \\ \hat{L}_{mi} = \sqrt{\hat{L}_{si}^2 - \frac{\hat{R}_{si} \hat{R}_{ri}}{a_{0i}}}. \end{cases} \quad (28)$$

In the numerical solution it-is considered that stator and rotor inductances have the same values in each winding.

SIMULATION RESULTS

Simulations have been performed to evaluate the proposed method. The machine model given by 1 was discretized by Euler technique under frequency of f_s = 5kHz. The SPIM was performed with a square wave reference of current and standstill rotor. The SPIM used is a four-pole, 368W, 1610rpm, 220V/3.4A. The parameters of this motor obtained from classical no-load and locked rotor tests are given in Table 1.

Table 1: Motor parameter obtained from classical tests.

R_{sq}	R_{rq}	L_{mq}	L_{sq}
7.00Ω	12.26Ω	0.2145H	0.2459H
R_{sd}	R_{rd}	L_{md}	L_{sd}
20.63Ω	28.01Ω	0.3370H	0.4264H

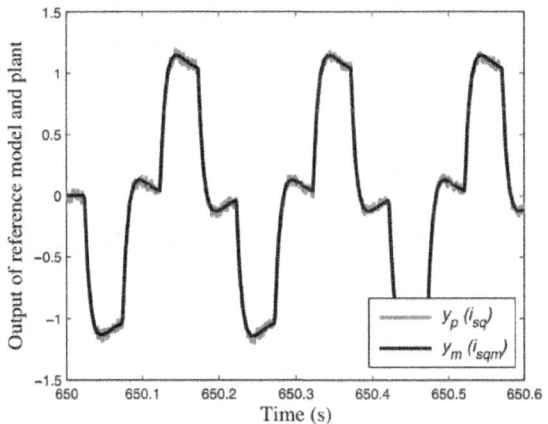

Figure 5: Plant and reference model output.

The reference model $W_m(s)$ is chosen so that the dynamic will be faster than plant output i_{sq}. Thus, the reference model is given by

$$W_m(s) = 180 \frac{s + 45}{s^2 + 180s + 8100}, \quad (29)$$

The induction motor is started in accordance with Fig. 2 with a Persistent Excitant reference current signal. A random noise was simulated to give nearly experimental conditions. Fig. 5 show the plant and reference model output after convergence of gains.

Fig. 6 shows the convergence of controller gains for parameter identification of q axis. This figure shows that gains reach a final value after 600s, demonstrating that parameter identification is possible. The gain convergence of d axis is shown in Fig. 7. Table 2 presents the final value of controller gains for the q and d axes, respectively.

Table 2: Final value of controller gains obtained in simulation.

θ_{1q}	θ_{2q}	θ_{3q}	θ_{4q}
-0.0096	-1.0925	0.8332	-0.0950
θ_{1d}	θ_{2d}	θ_{3d}	θ_{4d}
-0.0164	-0.6429	0.6885	-0.0352

The parameters of SPIM are obtained by combining the final value of controller gains from Table 2 with the equations 27, 28 and the reference model coefficients previously defined in equation 29. The results are shown in Table

3. It is possible to observe in simulation that the electrical parameters converge to machine parameters, even with noise in the currents.

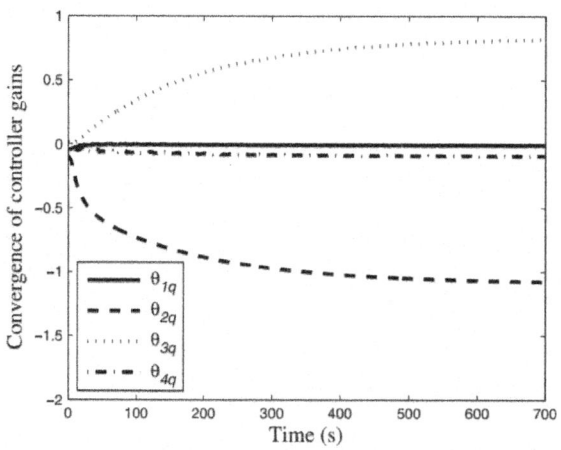

Figure 6: Convergence of controller gains vector for q axis.

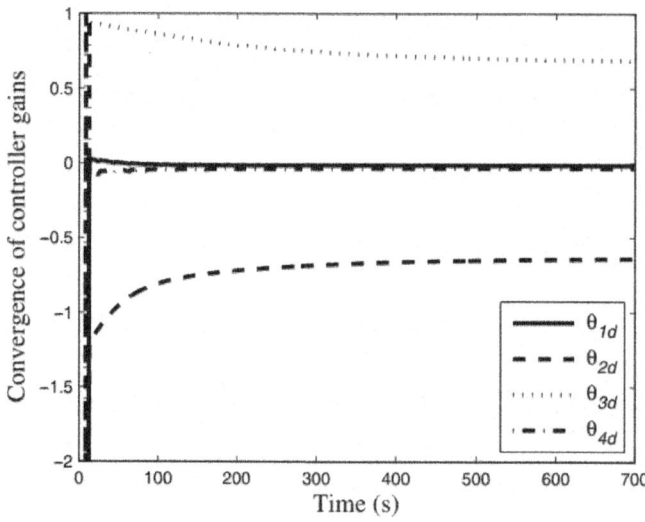

Figure 7: Convergence of controller gains vector for d axis.

Table 3: Motor parameter identified in simulation.

R_{sq}	R_{rq}	L_{mq}	L_{sq}
7.07Ω	12.21Ω	0.2150H	0.2462H
R_{sd}	R_{rd}	L_{md}	L_{sd}
20.22Ω	27.67Ω	0.3312H	0.4292H

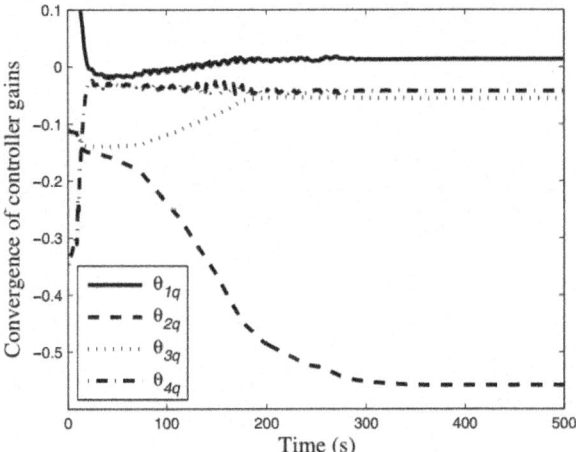

Figure 8: Experimental convergence of controller gains for q axis.

EXPERIMENTAL RESULTS

This section presents experimental results obtained from the induction motor described in simulation, whose electrical parameters obtained by classical no-load and locked rotor tests are shown in Table 1. The drive system consists of a three-phase inverter controlled by a TMS320F2812 DSP controller. The sampling period is the same used in the preceding simulation.

Unlike the simulation, unmodeled dynamics by drive, sensors and filters, among others, are included in the implementation. This implies that the plant model is a little different from physical plant. As a result, there is a small error that is proportional to plant uncertainties and is defined here as a residual error. The tracking error e_1 can be minimized by increasing the gradient gain P. However, increasing P in order to eliminate the residual error can cause divergence of controller gains and the system becomes unstable.

To overcome this problem a stopping condition was defined for the gain convergence. The identified stator resistance $\hat{R}_{si}$ was compared to measured stator resistance R_{si} obtained from measurements. Thus, the gradient gain P must be adjusted until the identified stator resistance is equal to the stator resistance measurement.

Figure 8 presents the convergence of controller gains for q axis. The gains reach a final value after 400s. Figure 9 presents the convergence of controller gains for d axis. The value of the gain that resulted in $\hat{R}_{si} = R_{si}$ was $\mathbf{P} = 20\mathbf{I}$.

The plant output i_{sq} and reference model output i_{sqm} are shown in Fig. 10, after controller gain convergence, where it is possible to see the residual error between the two curves. The final values of controller gains, for axes q and d, are shown in Table 4.

The parameters of SPIM are obtained by combining the final value of controller gains of Table 4 with the equations 27, 28 and the reference model coefficients previously defined in equation 29. The results are shown in Table 5.

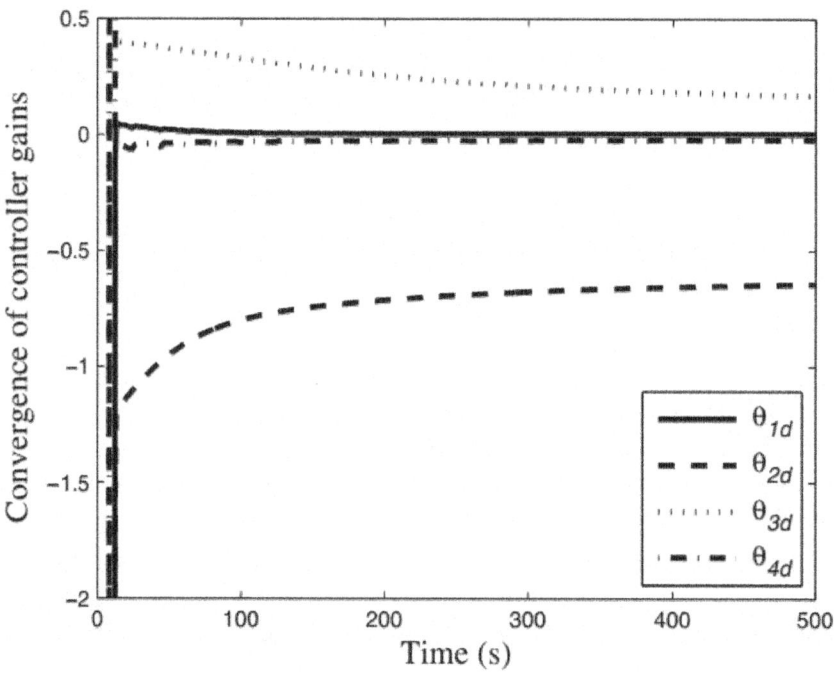

Figure 9: Experimental convergence of controller gains for d axis.

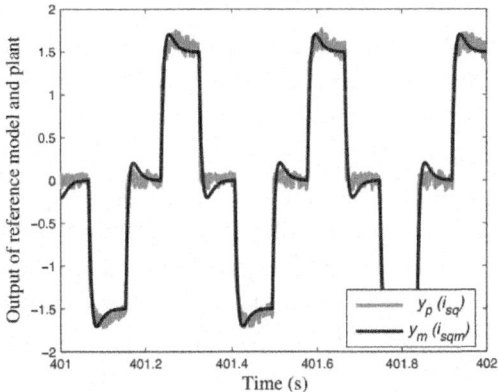

Figure 10: Experimental plant and reference model output.

Table 4: Final value of controller gains obtained in experimentation.

θ_{1q}	θ_{2q}	θ_{3q}	θ_{4q}
0.0136	-0.558	-0.0555	-0.0423
θ_{1d}	θ_{2d}	θ_{3d}	θ_{4d}
0.0042	-0.6345	0.1560	-0.0207

Table 5: Motor parameter identified in experimentation.

R_{sq}	R_{rq}	L_{mq}	L_{sq}
6.9105Ω	15.4181Ω	0.1821H	0.2593H
R_{sd}	R_{rd}	L_{md}	L_{sd}
20.9438Ω	34.9016Ω	0.4926H	0.6447H

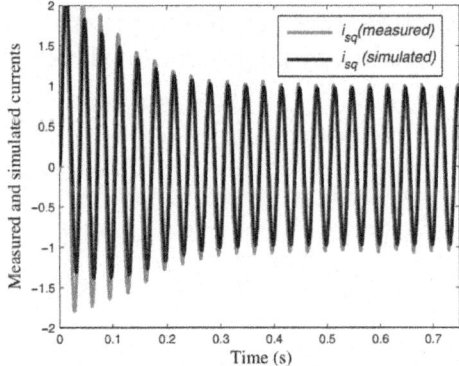

Figure 11: Transient of comparison among measured and simulated currents from q axis.

Comparison for Parameter Validation

A comparative study was performed to validate the parameters obtained by the proposed technique. The physical SPIM was fed by steps of 50V and 30Hz. The currents of main winding, auxiliary winding and speed rotor were measured.

Then, the dynamic model of SPIM given by 1 was simulated, using the parameters from Table 5, under the same conditions experimentally, i.e., steps of 50V and 30Hz. The measured rotor speed was used in the simulation model to make it independent of mechanical parameters. Thus, the simulated currents, from q and d windings, were compared with measured currents. Figures 11 and 12 shows the transient currents from axis q and d, respectively, while Figures 13 and 14 shows the steady-state currents from axis q and d, respectively.

From Figures 11-14, it is clear that the simulated machine with the proposed parameters presents similar behavior to the physical machine, both in transient and steady-state.

CONCLUSIONS

This chapter describes a method for the determination of electrical parameters of single phase induction machines based on a RMRAC algorithm, which initially was used in three-phase induction motor estimation in (Azzolin & Gründling, 2009). Using this methodology, it is possible to obtain all electrical parameters of SPIM for the simulation and design of an high performance control and sensorless SPIM drives. The main contribution of this proposed work is the development of automated method to obtain all electric parameters of the induction machines without the requirement of any previous test and derivative filters. Simulation results demonstrate the convergence of the parameters to ideal values, even in the presence of noise. Experimental results show that the parameters converge to different values in relation to the classical tests shown in Table 1. However, the results presented in Figures 11-14 show that the parameters obtained by proposed method present equivalent behavior to physical machine.

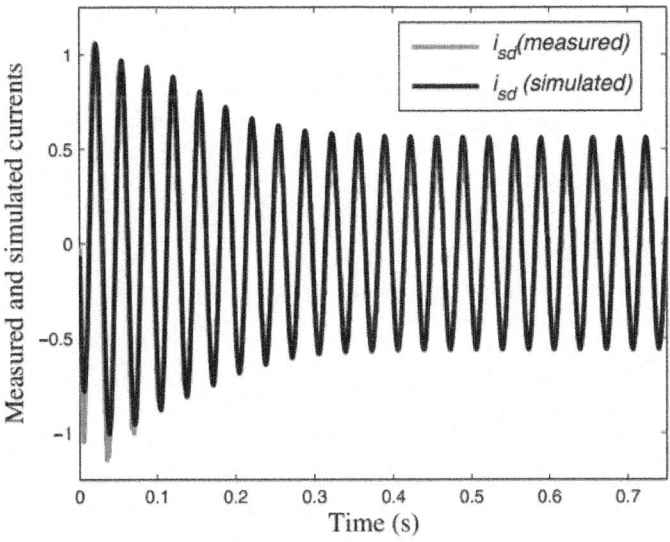

Figure 12: Transient of comparison among measured and simulated currents from d axis.

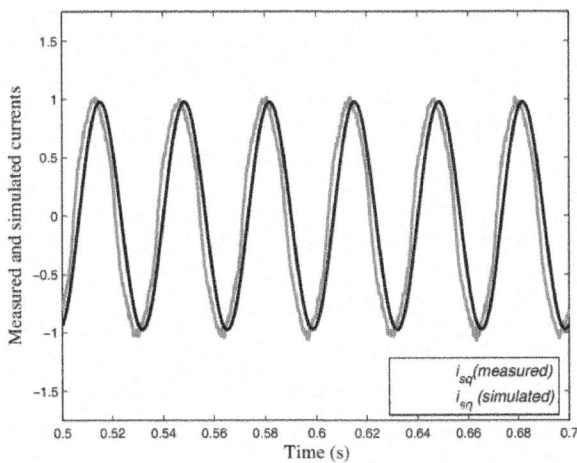

Figure 13: Steady-state of comparison among measured and simulated currents from q axis.

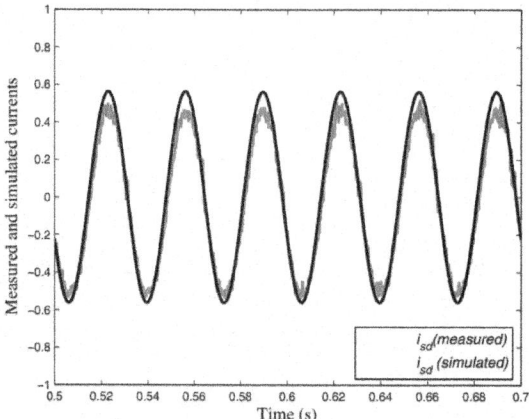

Figure 14: Steady-state of comparison among measured and simulated currents from d axis.

REFERENCES

1. Azzolin, R. & Gründling, H. (2009). A mrac parameter identification algorithm for three-phase induction motors, Electric Machines and Drives Conference, 2009. IEMDC '09. IEEE International, pp. 273 –278.
2. Azzolin, R., Martins, M., Michels, L. & Gründling, H. (2007). Parameter estimator of an induction motor at standstill, Industrial Electronics, 2009. IECON '09. 35th Annual Conference of IEEE, pp. 152 – 157.
3. Blaabjerg, F., Lungeanu, F., Skaug, K. & Tonnes, M. (2004). Two-phase induction motor drives, Industry Applications Magazine, IEEE 10(4): 24 – 32.
4. Câmara, H. & Gründling, H. (2004). A rmrac applied to speed control of an induction motor without shaft encoder, Decision and Control, 2004. CDC. 43rd IEEE Conference on, Vol. 4, pp. 4429 – 4434 Vol.4.
5. de Rossiter Correa, M., Jacobina, C., Lima, A. & da Silva, E. (2000). Rotor-flux-oriented control of a single-phase induction motor drive, Industrial Electronics, IEEE Transactions on 47(4): 832 –841.
6. Donlon, J., Achhammer, J., Iwamoto, H. & Iwasaki, M. (2002). Power modules for appliance motor control, Industry Applications Magazine, IEEE 8(4): 26 –34.
7. Ioannou, P. & Sun, J. (n.d.). Robust Adaptive Control, Prentice Hall. Ioannou, P. & Tsakalis, K. (1986). A robust direct adaptive controller,

Automatic Control, IEEE Transactions on 31(11): 1033 – 1043.
8. Koubaa, Y. (2004). Recursive identification of induction motor parameters, Simulation Modelling Practice and Theory 12(5): 363 – 381. URL: http://www.sciencedirect.com/science/article/B6X3C-4CJVR4N-1/2/ bcc4bdf719a9c7 ad5b99750423f5ff23
9. Krause, P., Wasynczuk, O. & Sudhoff, S. (n.d.). Analysis of Electric Machinery, NJ: IEEE Press.
10. Lozano-Leal, R., Collado, J. & Mondie, S. (1990). Model reference robust adaptive control without a priori knowledge of the high frequency gain, Automatic Control, IEEE Transactions on 35(1): 71 –78.
11. Ojo, O. & Omozusi, O. (2001). Parameter estimation of single-phase induction machines, Industry Applications Conference, 2001. Thirty-Sixth IAS Annual Meeting. Conference Record of the 2001 IEEE.
12. Ribeiro, L., Jacobina, C. & Lima, A. (1995). Dynamic estimation of the induction machine parameters and speed, Power Electronics Specialists Conference, 1995. PESC '95
13. Record., 26th Annual IEEE, Vol. 2, pp. 1281 –1287 vol.2. Vaez-Zadeh, S. & Reicy, S. (2005). Sensorless vector control of single-phase induction motor drives, Electrical Machines and Systems, 2005. ICEMS 2005. Proceedings of the Eighth International Conference on, Vol. 3, pp. 1838 –1842 Vol. 3.
14. Velez-Reyes, M., Minami, K. & Verghese, G. (1989). Recursive speed and parameter estimation for induction machines, Industry Applications Society Annual Meeting, 1989., Conference Record of the 1989 IEEE, pp. 607 –611 vol.1.
15. Vieira, R., Azzolin, R. & Gründling, H. (2009a). Parameter identification of a single-phase induction motor using rls algorithm, Power Electronics Conference, 2009. COBEP '09. Brazilian.
16. Vieira, R., Azzolin, R. & Gründling, H. (2009b). A sensorless single-phase induction motor drive with a mrac controller, Industrial Electronics, 2009. IECON '09. 35th Annual Conference of IEEE, pp. 1003 –1008.

Chapter 10

SWARM INTELLIGENCE BASED CONTROLLER FOR ELECTRIC MACHINES AND HYBRID ELECTRIC VEHICLES APPLICATIONS

Omar Hegazy[1], Amr Amin[2], and Joeri Van Mierlo[1]

[1]Faculty of Engineering Sciences Department of ETEC- Vrije Universiteit Brussel, Belgium
[2]Power and Electrical Machines Department, Faculty of Engineering – Helwan University, Egypt

INTRODUCTION

Swarm Intelligence in the form of Particle Swarm Optimization (PSO) has potential applications in electric drives. The excellent characteristics of PSO may be successfully used to optimize the performance of electric machines and electric drives in many aspects. It is estimated that, electric machines consume more than 50% of the world electric energy generated. Improving efficiency in electric drives is important, mainly, for two reasons: economic saving and reduction of environmental pollution. Induction motors have a high efficiency at rated speed and torque. However, at light loads, the iron losses increase dramatically, reducing considerably the efficiency. Swarm intelligence is used to optimize the performance of three applications; these applications are represented as follows:

- Losses Minimization of two asymmetrical windings induction motor
- Maximum efficiency and minimum operating cost of three-phase induction motor
- Optimal electric drive system for fuel cell hybrid electric vehicles.

In this chapter, a field-oriented controller that is based on Particle Swarm Optimization is presented. In this system, the speed control of two asymmetrical windings induction motor is achieved while maintaining maximum efficiency of the motor. PSO selects the optimal rotor flux level at each operating point. In addition, the electromagnetic torque is also improved while maintaining a fast dynamic response. A novel approach is used to evaluate the optimal rotor flux level by using Particle Swarm Optimization. PSO method is a member of the wide category of Swarm Intelligence methods (SI). The swarm intelligence is based on real life observations of social animals (usually insects), it is more flexibility and robust than any traditional optimization methods. PSO algorithm searches for global optimization for nonlinear problems with multi-objective. There are two speed control strategies explained in the next sections. These are field-oriented controller (FOC), and FOC based on PSO. The strategies are implemented mathematically and experimental. The simulation and experimental results have demonstrated that the FOC based on PSO method saves more energy than the conventional FOC method.

In this chapter, another application of PSO for losses and operating cost minimization control is presented for the induction motor drives. Two strategies for induction motor speed control are proposed. These strategies are maximum efficiency strategy (MES), based PSO, and minimum operating cost Strategy. The proposed technique is based on the principle that the flux level in a machine can be adjusted to give the minimum amount of losses and minimum operating cost for a given value of speed and load torque.

In the demonstrated systems, the powertrain components sizing and the power control strategy are the only adjustable parameters to achieve optimal power sharing between sources and optimal design with minimum cost, minimum fuel consumption, and maximum efficiency for Electric Vehicles (EVs) and Hybrid Electric Vehicles (HEVs). Their selection greatly influences the performance of the drive system in Hybrid Electric Vehicles applications. In this section, the design and power management control are investigated and optimized by using Particle Swarm Optimization.

LOSSES MINIMIZATION OF TWO ASYMMETRICAL WINDINGS INDUCTION MOTOR

In this section, a field orientation based on Particle Swarm Optimization (PSO) is applied to control the speed of two-asymmetrical windings induction motor. The maximum efficiency of the motor is obtained by the evaluation of optimal rotor flux at each operating point. In addition, the electro-magnetic torque is also improved while maintaining a fast dynamic response. In this section, a novel approach is used to evaluate the optimal rotor flux level. This approach is based on Particle Swarm Optimization (PSO). This section presents two speed control strategies. These are field-oriented controller (FOC) and FOC based on PSO. The strategies are implemented mathematically and experimental. The simulation and experimental results have demonstrated that the FOC based on PSO method saves more energy than the conventional FOC method [Hegazy, 2006; Amin et al., 2007; Amin et al., 2009]. The two asymmetrical windings induction motor is treated as a two-phase induction motor (TPIM). It is used in many low power applications, where three–phase supply is not readily available. This type of motor runs at an efficiency range of 50% to 65% at rated operating conditions.

The conventional field-oriented controller normally operates at rated flux at any values with its torque range. When the load is reduced considerably, the core losses become so high causing poor efficiency. If significant energy savings are required, it is necessary to optimize the efficiency of the motor. The optimum efficiency is obtained by the evaluation of the optimal rotor flux level. This flux level is varied according to the torque and the speed of the operating point.

PSO is applied to evaluate the optimal flux. It has the straightforward goal of minimizing the total losses for a given load and speed. It is shown that the efficiency is reasonably close to optimal.

Mathematical Model of the Motor

The d-q model of an unsymmetrical windings induction motor in a stationary reference frame can be used for a dynamic analysis. This model can take in account the core losses. The d-q model as applied to TPIM is described in [Hegazy, 2006; Amin et al., 2009]. The equivalent circuit is shown in Fig. 1

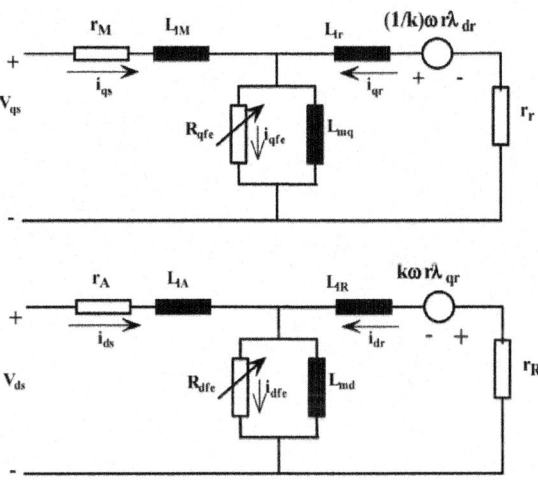

Figure 1: The d-q axes two-phase induction motor Equivalent circuit with iron losses.

The machine model may be expressed by the following voltage and flux equations: Voltage Equations (1):

$$v_{qs} = r_m i_{qs} + p\lambda_{qs} \tag{1}$$

$$v_{ds} = r_a i_{ds} + p\lambda_{ds} \tag{2}$$

$$0 = r_r i_{qr} - (1/k) * \omega_r \lambda_{dr} + p\lambda_{qr} \tag{3}$$

$$0 = r_R i_{ds} + k * \omega_r \lambda_{qr} + p\lambda_{dr} \tag{4}$$

$$0 = -i_{qfe} R_{qfe} + L_{mq}(p i_{qs} + p i_{qr} - p i_{qfe}) \tag{5}$$

$$0 = -i_{dfe} R_{dfe} + L_{md}(p i_{ds} + p i_{dr} - p i_{dfe}) \tag{6}$$

Flux Equations:

$$\lambda_{qs} = L_{lm} i_{qs} + L_{mq}(i_{qs} + i_{qr} - i_{qfe}) \tag{7}$$

$$\lambda_{ds} = L_{la} i_{ds} + L_{md}(i_{ds} + i_{dr} - i_{dfe}) \tag{8}$$

$$\lambda_{qr} = L_{lr} i_{qr} + L_{mq}(i_{qs} + i_{qr} - i_{qfe}) \tag{9}$$

$$\lambda_{dr} = L_{IR} i_{dr} + L_{md}(i_{ds} + i_{dr} - i_{dfe}) \tag{10}$$

Electrical torque equation is expressed as:

$$T_e = \frac{P}{2}(k L_{mq} i_{dr}(i_{qs} + i_{qr} - i_{qfe}) - \frac{1}{k} L_{md} i_{qr}(i_{ds} + i_{dr} - i_{qfe})) \tag{11}$$

$$T_e - T_l = j_m p\omega_r + B_m \omega_r \tag{12}$$

Field-Oriented Controller [FOC]

The stator windings of the motor are unbalanced. The machine parameters differ from the d axis to the q axis. The waveform of the electromagnetic torque demonstrates the unbalance of the system. The torque in equation (11) contains an AC term; it can be observed that two values are presented for the referred magnetizing inductance. It is possible to eliminate the AC term of electro-magnetic torque by an appropriate control of the stator currents. However, these relations are valid only in linear conditions. Furthermore, the model is implemented using a non-referred equivalent circuit, which presumes some complicated measurement of the magnetizing mutual inductance of the stator and the rotor circuits.

The indirect field-oriented control scheme is the most popular scheme for field-oriented controllers. It provides decoupling between the torque and the flux currents. The electric torque must be a function of the stator currents and rotor flux in synchronous reference frame [Popescu & Navrapescu, 2000]. Assuming that the stator currents can be imposed as:

$$i^s_{ds} = i^s_{ds1} \qquad (13)$$

$$i^s_{qs} = k\, i^s_{qs1} \qquad (14)$$

Where: $k = M_{srd} / M_{srq}$

$$T_e = \frac{P}{2L_r}\left[M_{sqr} i^s_{qs} \lambda^s_{dr} - M_{sdr} i^s_{ds} \lambda^s_{qr} \right] \qquad (15)$$

By substituting the variables i_{ds}, and i_{qs} by auxiliary variables i_{ds1}, and i_{qs1} into (15) the torque can be expressed by

$$T_e = \frac{P M_{sdr}}{2L_r}\left[i^s_{qs1} \lambda^s_{dr} - i^s_{ds1} \lambda^s_{qr} \right] \qquad (16)$$

In synchronous reference frame, the electromagnetic torque is expressed as:

$$T_e = \frac{P M_{sdr}}{2L_r}\left[i^e_{qs1} \lambda^e_{dr} - i^e_{ds1} \lambda^e_{qr} \right] \qquad (17)$$

$$T_e = \frac{P M_{sdr}}{2L_r}\left[i^e_{qs1} \lambda^e_{r} \right] \qquad (18)$$

$$i^e_{ds1} = \frac{\lambda^e_r}{M_{sdr}} \tag{19}$$

$$\omega_e - \omega_r = \frac{M_{sdr}}{\tau_r * \lambda_r} i^e_{qs1} \tag{20}$$

Synchronous reference frame for losses model

It is very complex to find the losses expression for the two asymmetrical windings induction motor with losses model. In this section, a simplified induction motor model with iron losses will be developed. For this purpose, it is necessary to transform all machine variables to the synchronous reference frame. The voltage equations are written in expanded form as follows [Hegazy, 2006; Amin et al., 2006; Amin et al., 2009]:

$$v^e_{qs} = r_m i^e_{qs} + L_{lm}\frac{di^e_{qs}}{dt} + L_{mq}\frac{di^e_{qm}}{dt} + \omega_e(L_{la}i^e_{ds} + L_{md}i^e_{dm}) \tag{21}$$

$$v^e_{ds} = r_a i^e_{ds} + L_{la}\frac{di^e_{ds}}{dt} + L_{md}\frac{di^e_{dm}}{dt} - \omega_e(L_{lm}i^e_{qs} + L_{mq}i^e_{qm}) \tag{22}$$

$$0 = r_r i^e_{qr} + L_{lr}\frac{di^e_{qr}}{dt} + L_{mq}\frac{di^e_{qm}}{dt} + \frac{\omega_{sl}}{k}(L_{lR}i^e_{dr} + L_{md}i^e_{dm}) \tag{23}$$

$$0 = r_R i^e_{dr} + L_{lR}\frac{di^e_{dr}}{dt} + L_{md}\frac{di^e_{dm}}{dt} - k*\omega_{sl}(L_{lr}i^e_{qr} + L_{mq}i^e_{qm}) \tag{24}$$

$$i^e_{qs} + i^e_{qr} = i^e_{qfe} + i^e_{qm} \tag{25}$$

$$i^e_{ds} + i^e_{dr} = i^e_{dfe} + i^e_{dm} \tag{26}$$

$$v^e_{dm} = -\frac{\omega_e L_{lr} L_{mqs}}{L_r} i^e_{qs} \tag{27}$$

$$v^e_{qm} = \omega_e L_{mds} i^e_{ds} \tag{28}$$

Where:

$$i^e_{dfe} = \frac{v^e_{qm}}{R_{qfe}}; \quad i^e_{dfe} = \frac{v^e_{dm}}{R_{dfe}}$$

The losses in the motor are mainly:
- Stator copper losses,
- Rotor copper losses,
- Core losses, and
- Friction losses

The total electrical losses can be expressed as follows

$$P_{losses} = P_{cu1} + P_{cu2} + P_{core} \qquad (29)$$

Where:

P_{cu1}: Stator copper losses

P_{cu2}: Rotor copper losses

P_{core}: Core losses

The stator copper losses of the two asymmetrical windings induction motor are caused by electric currents flowing through the stator windings. The core losses of the motor are produced from the hysteresis and eddy currents in the stator. The total electrical losses of motor can be rewritten as:

$$P_{losses} = r_m i_{qs}^2 + r_a i_{ds}^2 + r_r i_{qr}^2 + r_R i_{dr}^2 + \frac{v_{qm}^2}{R_{qfe}} + \frac{v_{dm}^2}{R_{dfe}} \qquad (30)$$

The total electrical losses are obtained as follows:

$$P_{losses} = \left[r_m + \frac{r_r L_{mqs}^2}{L_r^2} + \frac{\omega_e^2 L_{lr}^2 L_{mqs}^2}{L_r^2 R_{dfe}} \right] \frac{T_e^2 L_r^2}{p^2 \left(\frac{L_{mds}}{K} \right)^2 \lambda_r^2} + \left(r_a + \frac{\omega_e^2 L_{mds}^2}{R_{qfe}} \right) \frac{\lambda_r^2}{L_{mds}^2} \qquad (31)$$

$$\omega_{sl} = \frac{2 T_e * r_r}{P * \lambda_r^2} \qquad (32)$$

Where:

$\omega_e = \omega_r + \omega_{sl}$, and ω_{sl} is the slip speed (rad/sec).

Equation (31) is the electrical losses formula, which depends on rotor flux (λ_r) according to operating point (speed and load torque). The total losses of the motor (TP_{losses}) are given as follows:

$$TP_{losses} = P_{losses} + P_{Fric} = P_{in} - P_{out} \qquad (33)$$

$$\text{Efficiency } (\eta) = \frac{P_{out}}{P_{out} + TP_{losses}} \qquad (34)$$

Where:

Friction power losses $= F * \omega_r^2$, and Output power (Pout) $= T_L * \omega_r$.

Losses Minimization Control Scheme

The equation (33) is the cost function, the total losses, which depends on rotor flux (λ_r) according to the operating point. Figure 2 presents the distribution of losses in motor and its variation with the flux. As the flux reduces from the rated value, the core losses decrease, but the motor copper losses increase. However, the total losses decrease to a minimum value and then increase again. It is desirable to set the rotor flux at the optimal value, so that the efficiency is optimum [Hegazy, 2006; Amin et al., 2006; Amin et al., 2009].

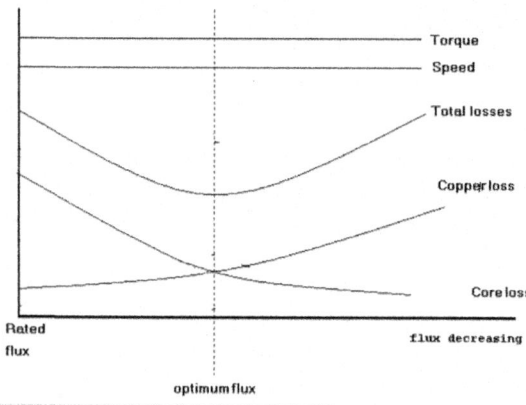

Figure 2: Losses variation of the motor with varying flux

PSO is applied to evaluate the optimal flux that minimizes the motor losses. The problem can be formulated as follows:

$$\text{minimize TP}_{losses} = \Gamma\,(\lambda r, T_L, \omega_r) \tag{35}$$

Particle Swarm Optimization (PSO)

Particle swarm optimization (PSO) was originally designed and introduced by Eberhart and Kennedy [Ebarhart, Kennedy, 1995; Ebarhart, Kennedy, 2001]. Particle Swarm Optimization (PSO) is an evolutionary computation technique (a search method based on a nature system). It can be used to solve a wide range of optimization problems. Most of the problems that can be solved using Genetic Algorithms (GA) could be solved by PSO. For example, neural network training and nonlinear optimization problems with continuous variables can be easily achieved by PSO [Ebarhart, Kennedy, 2001]. It can be easily expanded to treat problems with discrete variables.

The system initially has a population of random solutions. Each potential solution, called a particle. Each particle is given a random velocity and is flown through the problem space. The particles have memory and each particle keeps track of its previous best position (call the pbest) and with its corresponding fitness. There exit a number of pbest for the respective particles in the swarm and the particle with greatest fitness is called the global best (gbest) of the swarm. PSO can be represented by the concept of velocity and position. The Velocity of each agent can be modified by the following equations: (36 & 38):

$$v^{k+1} = w\, v_i^k + c_1 r_1 * (pbest - s_i^k) + c_2 r_2 * (gbest - s_i^k)$$

(36)

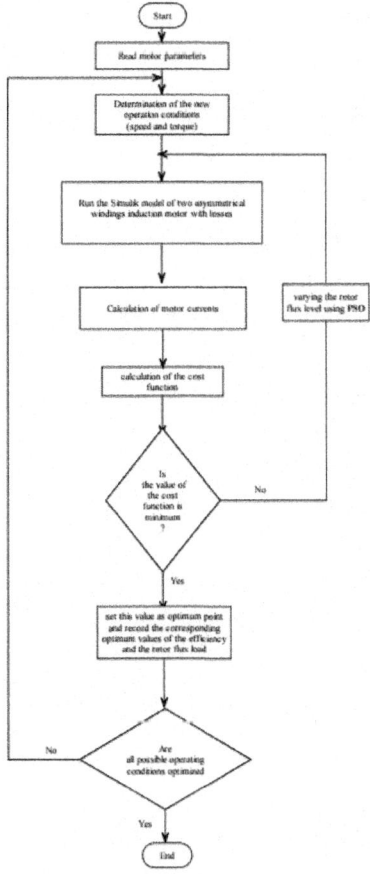

Figure 3: The flowchart of the execution of the PSO [Hegazy, 2006]

$$\omega = \omega_{max} - \frac{\omega_{max} - \omega_{min}}{iter_{max}} * iter$$

(37)

Using the above equations, a certain velocity can be calculated that gradually gets close to (pbest) and (gbest). The current position (searching point in the solution space) can be modified by the following equation:

$$S_i^{k+1} = S_i^k + v_i^{k+1} \qquad (38)$$

Where:

v^k : Current velocity of agent i at iteration.

v_i^{k+1} : Modified velocity of agent i

r_1, r_2 : random number distributed [0,1],

S_i^k : current position of agent i,

: weight function for velocity of agent i,

c_1, c_2 : positive constants; [c1+ c2< 4].

ω_{max} : Initial weight,

ω_{min} : Final weight,

$iter_{max}$: Maximum iteration number,

iter : Current iteration number.

- In (35), the losses formula is the cost function of the PSO. The particle swarm optimization (PSO) technique is used for minimizing this cost function.
- The PSO is applied to evaluate the optimal rotor flux that minimizes the motor losses at any operating point. Figure 3 presents the flowchart of the execution of PSO, which evaluates the optimal flux by using MATLAB /SIMULINK.

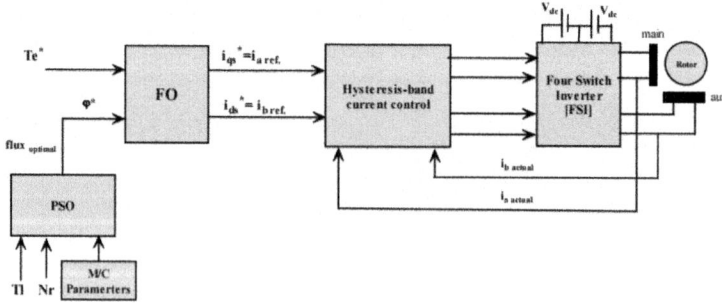

Figure 4: The proposed losses minimization control system.

The optimal flux is the input of the indirect rotor flux oriented controller. The indirect fieldoriented controller generates the required two reference currents to drive the motor corresponding to the optimal flux. These currents are fed to the hysteresis current controller of the two-level inverter. The switching pattern is generated according to the difference between the reference current and the load current through the hysteresis band. Figure 4 shows a whole control diagram of the proposed losses-minimization control system.

Simulation Results

In this section, the proposed application is implemented numerically using MATLABSIMULINK to validate the performance of the proposed control strategy. The motor used in this study has the following parameters, which were measured by using experimental tests. Table 1 shows motor parameters. The used parameters of the PSO are shown in Table 2.

Table 1: Motor Parameters

Rated power	750 w
V	220 v
F	50 Hz
r_M	4.6 Ω
r_A	10.6 Ω
X_{Lm}	4.31 Ω
X_{La}	7.1472 Ω
rr	3.455 Ω
X_{Lr}	4.284 Ω
X_{mq}	89.65Ω
X_{md}	169.43Ω
R_{qfe}	1050Ω
R_{dfe}	1450Ω
J	0.005776 kg.m2
B	0.00328N.m.sec/r
Pole pair	2

The optimal rotor flux provides the maximum efficiency at any operating point. There are six-cases of the motor operation are studied by using FOC based on PSO. PSO will evaluate the optimal rotor flux level. This flux is fed to the FOC module. Figure 5 shows the performance of the motor at case (1) (T_L=0.25 PU, N_r=0.5 N_{rated}), when PSO is applied side-by-side FOC as shown in Fig.4.

Table 2: PSO Algorithm parameter

Population size	10
Max. iter	50
c1	0.5
c2	0.5
Max. weight	1.4
Min. weight	0.1
r1	[0,1]
r2	[0,1]
Lower Bound	0.2
Upper Bound	2

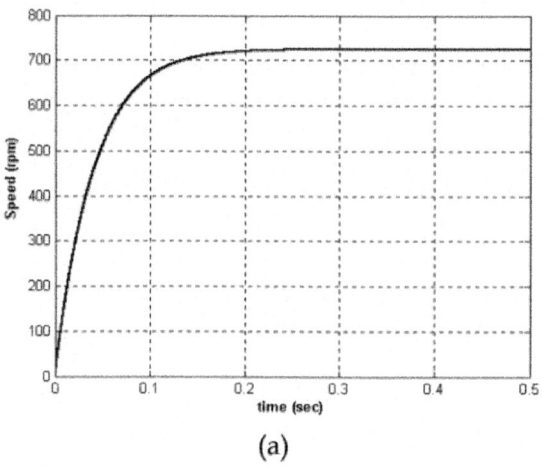

(a)

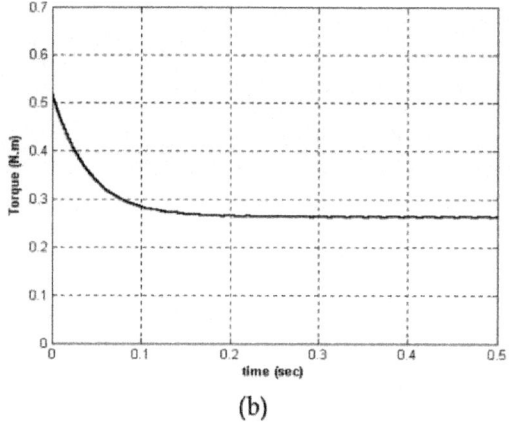

(b)

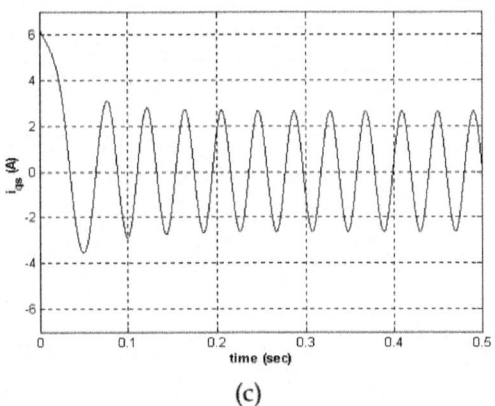

(c)

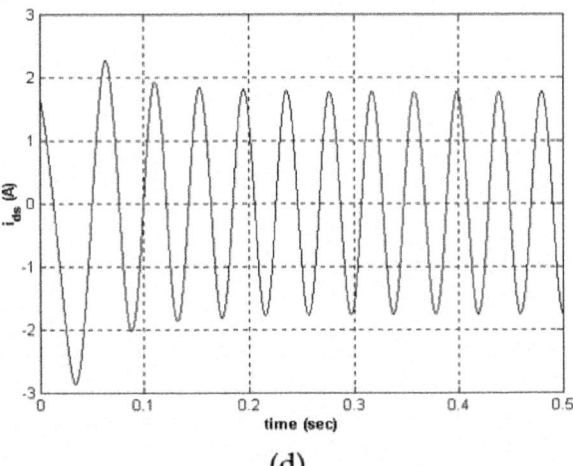

(d)

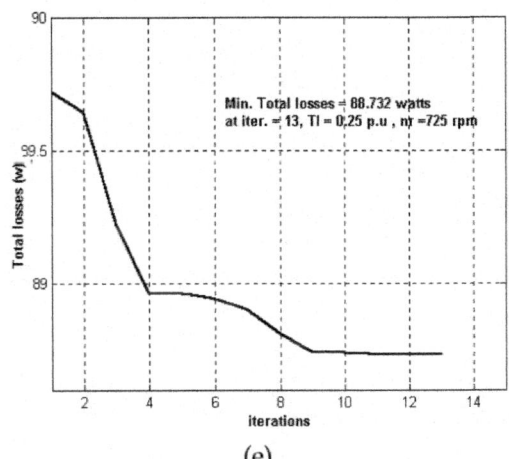

(e)

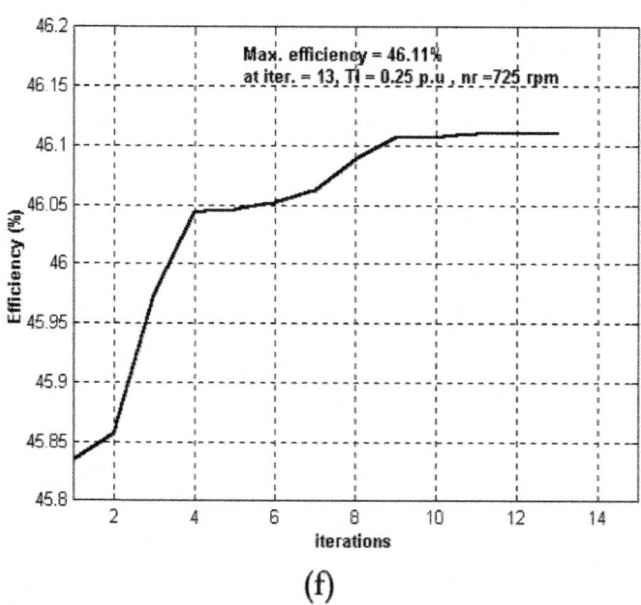

(f)

Figure 5: Simulation results of the motor at case (1).Speed-time curve, (b) Torque-time curve, (c) The stator current in q-axis, (d) The stator current in d-axis, (e) Total Losses against iterations, (f) Efficiency against iterations

Figure 6 illustrates the comparison between FOC and FOC based PSO control methods at different operating points. Figure 7 presents the optimal flux, which is obtained by applying PSO. Table 3 presents the summary of the results of FOC and FOC based PSO methods.

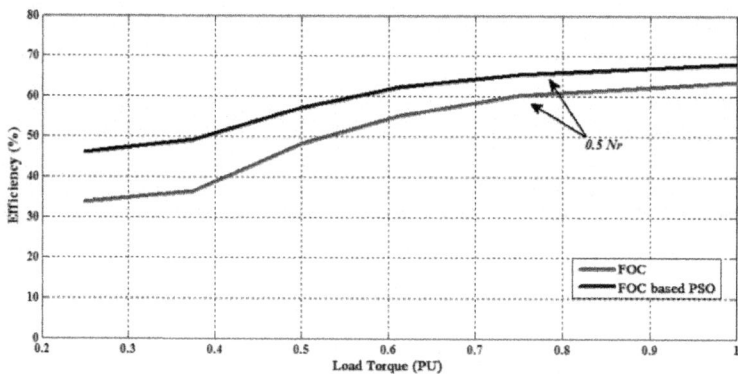

Figure 6: the comparison between FOC and FOC based PSO

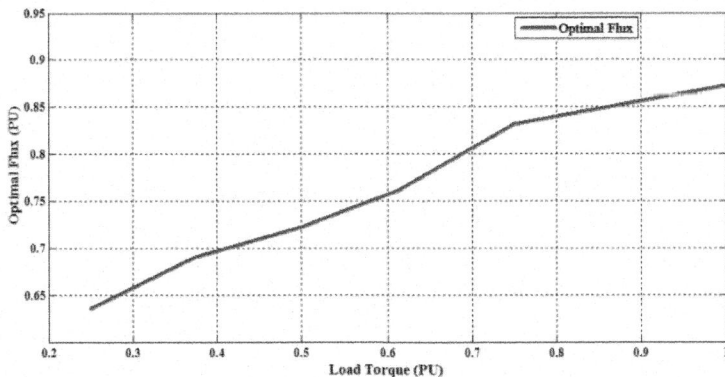

Figure 7: The optimal flux at different load torque

Table 3: Summary of the results of the two controllers

Cases	T_L (PU)	N_r (rpm)	FOC		FOC based PSO		Improvement (%)
			λ (PU)	η (%)	$\lambda_{Optimal}$ (PU)	η (%)	
(1)	0.25	0.5 N_{rated}	1	33.85	0.636	46.11	36.22
(2)	0.375	0.5 N_{rated}	1	36.51	0.6906	49.15	34.62
(3)	0.5	0.5 N_{rated}	1	48.21	0.722	57.11	18.46
(4)	0.6125	0.5 N_{rated}	1	55.15	0.761	62.34	13.04
(5)	0.75	0.5 N_{rated}	1	60.175	0.8312	65.31	8.53
(6)	1	0.5 N_{rated}	1	63.54	0.8722	68.15	7.26

In practical system, the flux level based on PSO at different operating points (torque and speed) is calculated and stored in a look up table (LUT). The use of look up table will enable the system to work in real time without any delay that might be needed to calculate the optimal point. The proposed controller would receive the operating point (torque and speed) and get the optimum flux ($\lambda_{optimal}$) from the look up table. It will generate the required reference current. It is noticed that, the efficiency with the FOC based on PSO method is higher than the efficiency with the FOC method only.

Experimental Results

To verify the validity of the proposed control scheme, a laboratory prototype is built and tested [Hegazy, 2006; Amin et al., 2006; Amin et al., 2009]. The basic elements of the proposed experimental scheme are shown in Fig. 8 and Fig. 9. The experimental results of the motor are achieved by coupling the motor

to an eddy current dynamometer. The experimental results are achieved using two control methods:
- Field-Oriented Control [FOC], and
- Field-Oriented Control [FOC] based on PSO.

The reference and the actual motor currents are fed to the hysteresis current controller. The switching pattern of the two-level four-switch inverter [FSI] is generated according to the difference between the reference currents and the load currents. Figure 10 shows the experimental results of the motor with FOC at case (1), where the motor is loaded by $T_L = 0.25$ PU. Figure 11 shows the experimental result of the motor with FOC based on PSO at case (1). The cases are summarized in Table 4.

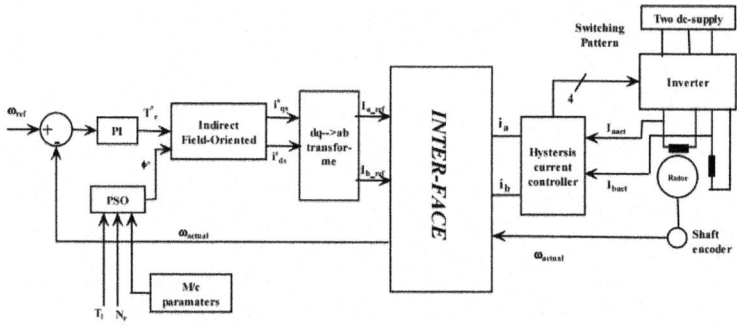

Figure 8: Block diagram of the proposed experimental scheme [Hegazy, 2006; Amin et al., 2009]

Table 4: The summary of the two-cases

Cases	FOC			FOC with PSO			Improvement (%)
	λ (PU)	Power Input (W)	η (%)	λ (PU)	Power Input (W)	η (%)	
(1)	1	235	32.3	0.636	169	44.92	39.07
(2)	1	323	35.2	0.690	243	47.06	33.69

Swarm Intelligence Based Controller For Electric Machines... 217

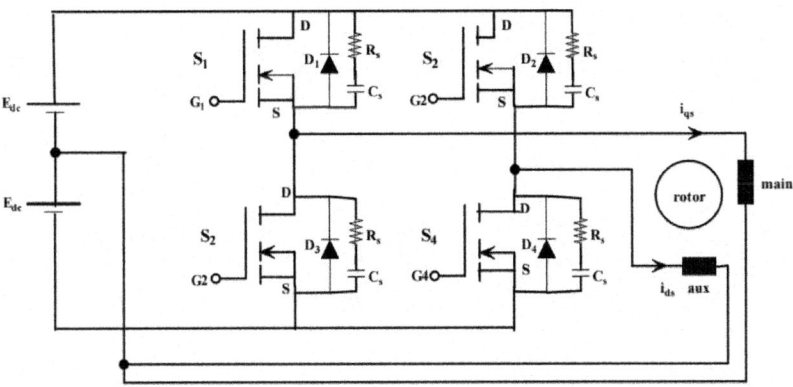

Figure 9: The power circuit of Four Switch Inverter [FSI]

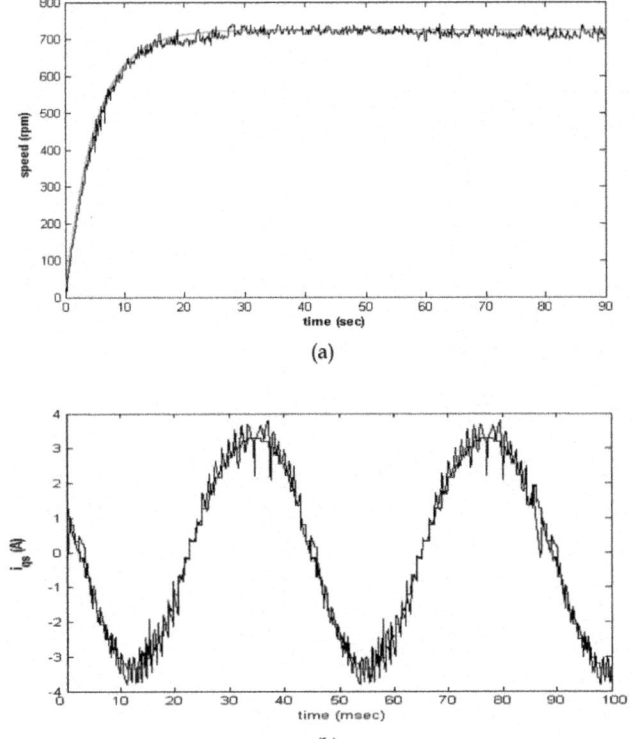

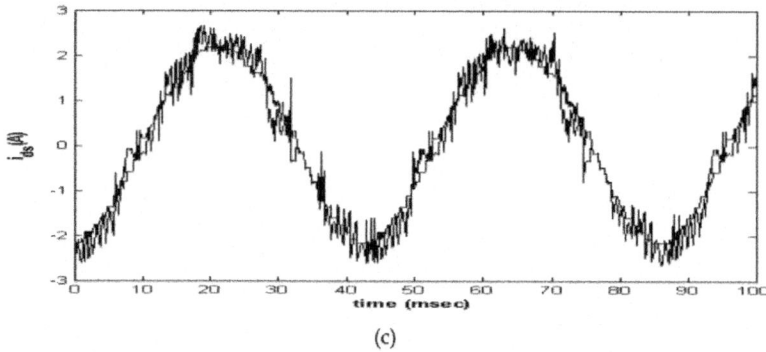

(c)

Figure 10: Experimental results of FOC method; the reference and actual speed, (b) the reference and actual current in q-axis, (c) The reference and actual current in d-axis

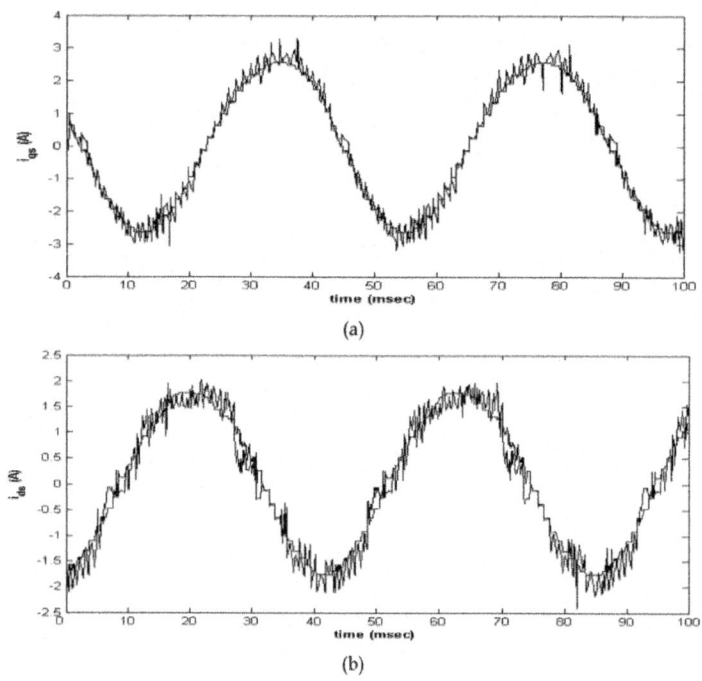

(a)

(b)

Figure 11: Experimental results of FOC method based on PSO. (a) The reference and actual current in q-axis, the reference and actual current in d-axis

Finally, these results demonstrate that, the FOC based on PSO method saves more energy than conventional FOC method. Thus, the efficiency with PSO is improved than it's at FOC.

MAXIMUM EFFICIENCY AND MINIMUM OPERATING COST OF INDUCTION MOTORS

This section presents another application of PSO for losses and operating cost minimization control in the induction motor drives. Two control strategies for induction motor speed control are proposed. Those two strategies are based on PSO and called Maximum Efficiency Strategy and Minimum Operating Cost Strategy [A. Hamid et al. 2006]. The proposed technique is based on the principle that the flux level in the machine can be adjusted to give the minimum amount of losses and minimum operating cost for a given value of speed and load torque. The main advantages of the proposed technique are; its simple structure. It is a straightforward maximization of induction motor efficiency and its operating cost for a given load torque. As was demonstrated, PSO is efficient in finding the optimum operating machine's flux level. The optimum flux level is a function of the machine operating point.

The main induction motor losses are usually split into five components: stator copper losses, rotor copper losses, iron losses, mechanical losses, and stray losses [Kioskeridis & Margaris, 1996].

The efficiency that decreases with increasing losses can be improved by minimizing the losses. Copper losses reduce with decreasing the stator and the rotor currents, while the core losses essentially increase with increasing air-gap flux density. A study of the copper and core losses components reveals that their trends conflict. When the core losses increase, the copper losses tends to decrease. However, for a given load torque, there is an air-gap flux density at which the total losses is minimized. Hence, electrical losses minimization process ultimately comes down to the selection of the appropriate air-gap flux density of operation. Since the air-gap flux density must be variable when the load is changing, control schemes in which the (rotor, air-gap) flux linkage is constant will yield sub-optimal efficiency operation especially when the load is light. Then to improve the motor efficiency, the flux must be reduced when it operates under light load conditions by obtaining a balance between copper and iron losses.

The challenge to engineers, however, is to be able to predict the appropriate flux values at any operating points over the complete torque and speed range which will minimize the machines losses, hence maximizing the efficiency. In general, there are three different approaches to improve the induction motor efficiency especially under light-load.

Losses Model Controller (LMC)

This controller depends on a motor losses model to compute the optimum flux analytically. The main advantage of this approach is its simplicity and it does not require extra hardware. In addition, it provides smooth and fast adaptation of the flux, and may offer optimal performance during transient operation. However, the main problem of this approach is that it requires the exact values of machine parameters. These parameters include the core losses and the main inductance flux saturation, which are unknown to the users and change considerably with temperature, saturation, and skin effect. In addition, these parameters may vary due to changes in the operating conditions. However, with continuing improvement of evolutionary parameter determination algorithms, the disadvantages of motor parameters dependency are slowly disappearing.

Search Controller (SC)

This controller measures the input power of the machine drive regularly at fixed time intervals and searches for the flux value, which results in minimum power input for given values of speed and load torque. This particular method does not demand knowledge of the machine parameters and the search procedure is simple to implement.

However, some disadvantages appear in practice, such as continuous disturbances in the torque, slow adaptation (7sec.), difficulties in tuning the algorithm for a given application, and the need for precise load information. In addition, the precision of the measurements may be poor due to signal noise and disturbances. This in turn may cause the SC method to give undesirable control performance. Moreover, nominal flux is applied in transient state and is tuned after the system reaches steady state to an optimal value by numerous increments, thus lengthening the optimization process. Therefore, the SC technique may be slow in obtaining the optimal point. In addition, in real systems, it may not reach a steady state and so cause oscillations in the air gap flux that result in undesirable torque disturbances. For these reasons, this is not a good method in industrial drives.

Look Up Table Scheme

It gives the optimal flux level at different operating points. This table, however, requires costly and time-consuming prior measurements for each motor. In this section, a new control strategy uses the loss model controller based on PSO is proposed. This strategy is simple in structure and has the straightforward goal of maximizing the efficiency for a given load torque. The resulting induction motor efficiency is reasonably close to optimal. It is well known that the presence of uncertainties, the rotor resistance, for instance makes the result no more optimal. Digital computer simulation results are obtained to demonstrate the effectiveness of the proposed method.

Definition of Operating Strategies

The following definitions are useful in subsequent analyses. Referring to the analysis of the induction motor presented in [A. Hamid et al. 2006], the per-unit frequency is

$$a = \frac{\omega_e}{\omega_b} = \frac{\omega_s + \omega_r}{\omega_b} \tag{39}$$

The slip is defined by

$$s = \frac{\omega_s}{\omega_b} = \frac{\omega_s}{\omega_s + \omega_r} \tag{40}$$

The rotor current is given by

$$I_r' = \frac{\phi_m}{\sqrt{\left(\frac{r_r'}{sa}\right)^2 + X_{lr}'^2}} \tag{41}$$

The electromagnetic torque is given by

$$T_e = \frac{\left(\frac{r_r'}{sa}\right)}{\left(\frac{r_r'}{sa}\right)^2 + X_{lr}'^2} \phi_m^2 \tag{42}$$

The stator current is related to the air gap flux and the electromagnetic torque as:

$$I_s = \sqrt{\left(S_1 \phi_m + S_2 \phi_m^3 + S_2 \phi_m^5\right)^2 + C_L \frac{T_e^2}{\phi_m^2}} \tag{43}$$

Where

$$C_L = 1 + 2 \times \frac{X_{lr}'}{X_m}$$

The air gap flux is related to the electromagnetic torque as:

$$\phi_m = \sqrt{\frac{sa}{r_r'} \sqrt{\left(\frac{r_r'}{sa}\right)^2 + X_{lr}'^2}} \sqrt{T_e} \tag{44}$$

The efficiency is defined as the output power divided by the electric power supplied to the stator (inverter losses are included):

$$\text{Efficiency}(\eta) = \frac{P_{out}}{P_{in}} \tag{45}$$

Maximum Efficiency Strategy

In MES (Maximum Efficiency Strategy), the slip frequency is adjusted so that the efficiency of the induction motor drive system is maximized [A. Hamid et al. 2006].

The induction motor losses are the following:

1. Copper losses: these are due to flow of the electric current through the stator and rotor windings and are given by:

$$P_{cu} = r_s I_s^2 + r_r' I_r'^2 \qquad (46)$$

2. Iron losses: these are the losses due to eddy current and hysteresis, given by

$$P_{core} = K_e(1+S^2)a^2 \varphi_m^2 + K_h(1+S)a\varphi_m^2 \qquad (47)$$

3. Stray losses: these arise on the copper and iron of the motor and are given by

$$P_{cu} = C_{str} \omega_r^2 I_r'^2 \qquad (48)$$

4. Mechanical losses: these are due to the friction of the machine rotor with the bearings and are given by:

$$P_{fw} = C_{fw} + \omega_r^2 \qquad (49)$$

5. Inverter losses: The approximate inverter loss as a function of stator current is given by:

$$P_{inv} = K_{1inv} i_s^2 + K_{2inv} i_s \qquad (50)$$

Where: K_{1inv}, K_{2inv} are coefficients determined by the electrical characteristics of a switching element where: K1inv= 3.1307e-005, K2inv=0.0250.

The total power losses are expressed by:

$$P_{losses} = P_{cu} + P_{core} + P_s + P_{fw} + P_{inv} = \left[r_s I_s^2 + r_r' I_r'^2 \right] + \left[K_e(1+S^2)a^2 \varphi_m^2 \right] + \left[K_h(1+S)a\varphi_m^2 \right] + \left[C_{str} \omega_r^2 I_r'^2 \right] + \left[K_{1inv} i_s^2 + K_{2inv} i_s \right] \qquad (51)$$

The output power is given by:

$$P_{out} = T_L \times \omega_r \qquad (52)$$

The input power is given by:

$$P_{in}=P_{out}+P_{losses} = P_{cu} + P_{core} + P_s + P_{fw} + P_{inv} = \left[r_s I_s^2 + r_r' I_r'^2\right] +$$
$$\left[K_c(1+S^2)a^2\phi_m^2\right] + \left[K_h(1+S)a\phi_m^2\right] + \left[C_{str}\omega_r^2 I_r'^2\right]$$
$$+ \left[K_{1inv} i_s^2 + K_{2inv} i_s\right] + T_L \times \omega_r \tag{53}$$

The efficiency is expressed as:

$$\eta = \frac{T_L \times \omega_r}{\left[\left[r_s I_s^2 + r_r' I_r'^2\right] + \left[K_c(1+S^2)a^2\phi_m^2\right] + \left[K_h(1+S)a\phi_m^2\right] + \left[C_{sr}\omega_r^2 I_r'^2\right]\right.}{\left. + \left[K_{1inv} i_s^2 + K_{2inv} i_s\right] + T_L \times \omega\right]} \tag{54}$$

The efficiency maximization of the induction motor problem can be formulated as follows:

$$\text{Maximize} \quad \eta(T_L, \omega_s, \omega_r) \tag{55}$$

The maximization should observe the fact that the amplitude of the stator current and flux cannot exceed their specified maximum point.

Minimum Operating Cost Strategy

In Minimum Operating cost Strategy (MOCS), the slip frequency is adjusted so that the operating cost of the induction motor is minimized. The operating cost of the induction machine should be calculated over the whole life cycle of the machine. That calculation can be made to evaluate the cost of the consumed electrical energy. The value of average energy cost considering the power factor penalties can be determined by the following stages [A. Hamid et al. 2006]:

1. If $0 \leq PF < 0.7$

$$C = C_0\left[1 + \left(\frac{0.9 - PF}{0.01}\right) \times \frac{1}{100}\right] \tag{56}$$

2. If $0.7 \leq PF \leq 0.92$, If $PF \geq 0.9$, $PF = 0.9$

$$C = C_0\left[1 + \left(\frac{0.9 - PF}{0.01}\right) \times \frac{0.5}{100}\right] \tag{57}$$

3. If $0.9 \leq PF \leq 1$, If $0.95 \leq PF \leq 1$, $PF = 0.95$

$$C = C_0\left[1 + \left(\frac{0.9 - PF}{0.01}\right) \times \frac{0.7}{100}\right] \tag{58}$$

If the average energy cost C is calculated, it can be used to establish the present value of losses. The total cost of the machine is the sum of its initial cost plus the present worth value of losses and maintenance costs.

$$PW_L = C \times T \times N \times P_{out} \times \left[\frac{1}{\eta} - 1\right] \tag{59}$$

Where:

PW_L = present worth value of losses
C0   = energy cost per kWh,
        = modified energy cost per kWh
T   = running time per year (Hrs / year)
N   = evaluation life (years)
P_{out} = the output power (kW)
        = the efficiency

The operating cost minimization of the induction motor problem can be formulated as follows:

$$\text{Minimize } PW_L(T_L, \omega_s, \omega_r) \tag{60}$$

Simulation Results

The simulation is carried out on a three-phase, 380 V, 1-HP, 50 Hz, and 4-pole, squirrel cage induction motor. The motor parameters are R_s=0.0598, X_{ls}=0.0364, X_m=0.8564, X_{lr}=0.0546, R_r=0.0403, K_e=0.0380, K_h=0.0380, C_{str}=0.0150, C_{fw}=0.0093, S_1=1.07, S_2=-0.69, S_3=0.77. For cost analysis, the following values were assumed: C_0=0.05, N=15, T=8000. The task of PSO controller is to find that value of slip at which the maximum efficiency occurs. At certain load torque and rotor speed, the PSO controller determines the slip frequency ω_s at which the maximum efficiency and minimum operating cost occur. The block diagram of the optimization process based on PSO is shown in Fig.12. To observe the improvements in efficiency using the suggested PSO controller, Fig. 13 shows the efficiency of the selected machine for all operating conditions using conventional methods (constant voltage to frequency ratio, field oriented control strategy) and using the proposed PSO controller at different rotor speed levels, Wr = 0.2 PU, and Wr = 1 PU respectively [A. Hamid et al. 2006]. This figure shows that a considerable energy saving is achieved in comparison with the conventional method (field oriented control strategy and constant voltage to frequency ratio) especially at light loads and small rotor speed. Figure 14 compares the efficiency of the induction motor drive system under the maximum efficiency strategy with the minimum operating cost strategy at Wr = 1 PU. It is obvious from the figure that the efficiency is almost the same for both strategies for all operating points.

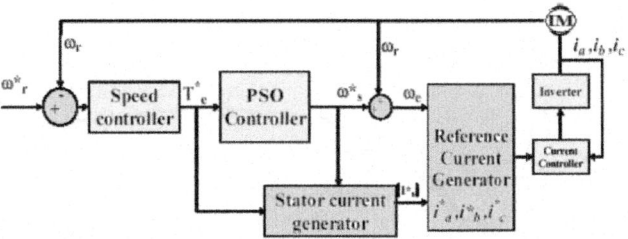

Figure 12: The proposed drive system based on PSO controller

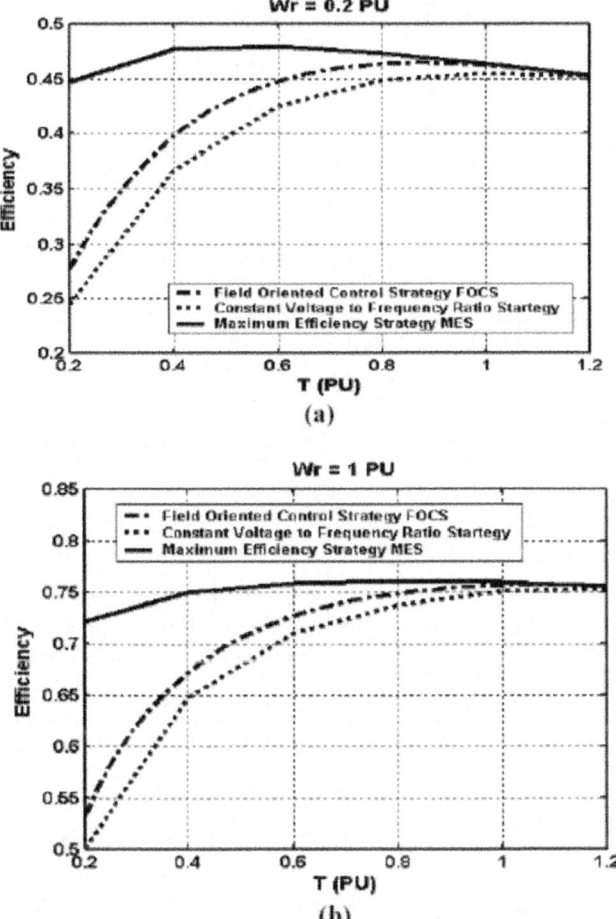

Figure 13: The efficiency of the induction motor using the maximum effi-

ciency strategy compared with the efficiency using the conventional methods at (a) Wr = 0.2 PU, (b) Wr= 1 PU [A. Hamid et al. 2006].

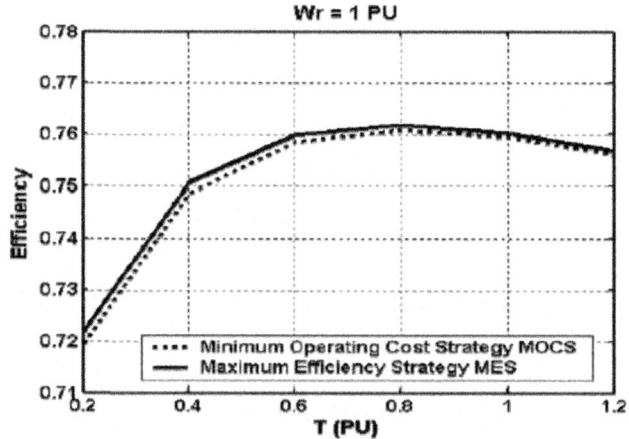

Figure 14: The efficiency of the induction motor using the maximum efficiency strategy compared with the efficiency using minimum operating cost strategy at Wr= 1 PU

Table 5 shows the efficiency comparison using few examples of operating points.

Table 5: Some examples of efficiency comparison under different Load torque levels and W_r = 1 PU [A. Hamid et al. 2006].

T (PU)	Efficiency comparison for ω_r = 1 PU			
	Constant voltage to frequency ratio	Field oriented control	Maximum efficiency strategy	Minimum Operating Cost Strategy
0.2	0.5003	0.5330	0.7217	0.7193
0.4	0.6482	0.6730	0.7506	0.7485
0.6	0.7100	0.7271	0.7598	0.7584
0.8	0.7384	0.7494	0.7618	0.7608
1	0.7508	0.7569	0.7603	0.7595
1.2	0.7544	0.7566	0.7568	0.7562

Figure 15 compares the power factor of the induction motor drive system under the maximum efficiency strategy with the minimum operating cost strategy at Wr = 1 PU. Finally, the proposed PSO-controller adaptively adjusts the slip frequency such that the drive system is operated at the minimum loss and minimum operating cost. It was found that the optimal system slip changes with variations in speed and load torque. When comparing the proposed strategy with the conventional methods field oriented control strategy and constant voltage to frequency ratio). It was found that a significant efficiency improvement especially at light loads for all speeds. On the other hand, small efficiency improvement is achieved at near rated loads (see Fig.13, and Fig.15).

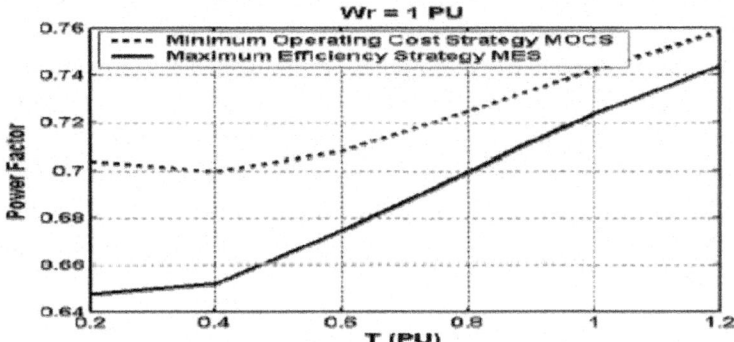

Figure 15: The power factor of the induction motor using the maximum efficiency strategy compared with the efficiency using minimum operating cost strategy at Wr= 1 PU [A. Hamid et al. 2006]

OPTIMAL ELECTRIC DRIVE SYSTEM FOR FUEL CELL HYBRID ELECTRIC VEHICLES

Although there are various FC technologies available for use in vehicular systems, the proton exchange membrane FC (PEMFC) has been found to be a prime candidate, since PEMFC has higher power density and lower operating temperatures when compared to the other types of FC systems. A stand-alone FC system integrated into an automotive powertrain is not always sufficient to satisfy the load demands of a vehicle. Although FC systems exhibit good power capability during steady-state operation, the response of fuel cells during transient and instantaneous peak power demands is relatively poor. Thus, the FC system can be hybridized with supercapacitors (SC) or batteries to meet the total power demand of a hybrid electric vehicle (HEV) [Van Mierlo et. al, 2006; Paladini et. al, 2007].

In this section, a new control strategy based PSO algorithm is proposed for the Fuel Cell/Supercapacitor hybrid electric vehicles to optimize the electric drive system [Hegazy & Van Mierlo, 2010]. Many factors influence on the performance of the electric drive system. These factors are mass, volume, size, efficiency, fuel consumption and control strategy. Therefore, the PSO is proposed to minimize the cost, the size and the mass of the powertrain sources (Fuel cell, and supercapacitor) as well as minimum fuel consumption and improves the efficiency of the system. PSO algorithm searches for global optimization for nonlinear problems with multi-objective. For a given driving cycle, the size and the cost of fuel cell and supercapacitor are minimized by identifying the best number of units of each, respectively. Three methods have been designed to achieve the optimal sizing. These are conventional method, trial and error, as was mentioned in [Wu & Gao, 2006], GA, and PSO. In addition, the hydrogen consumption is minimized by the evaluation of the optimal power distribution between fuel cell (main source) and supercapacitor (auxiliary source). Three control strategies are implemented to minimize the hydrogen consumption and maintain the state of charge (SOC) of the supercapacitor (SOCinitial =SOCfinal), which are control strategy based on Efficiency Map (CSEM), Control strategy based on PSO (CSPSO), and control strategy based on GA (CSGA).

System Description

The power system configuration is illustrated in Fig.16. A hybrid fuel cell/supercapacitor vehicle utilizes a PEM fuel cell as the main power source and a supercapacitor as the auxiliary power source. A multiple-input power electronic converter (MIPEC) is proposed to interface the traction drive requirements. In the MIPEC, the FC is connected to DC Bus via a Boost DC/DC converter ($\eta_B = \eta_{conv}$) and the supercapacitor is connected to DC Bus via a Buck/Boost converter ($\eta_{B/B} = \eta_{conv}$). The desired value of the DC-Bus voltage is chosen to be 400 V with variations of ± 10% are permissible. The power supplied by the powertrain has to be obtaining from the power demand predicted by the dynamics of the vehicle. The efficiency of each component in the hybrid powertrain is taken into account. A detailed model of the powertrain is built in MATLAB /SIMULINK.

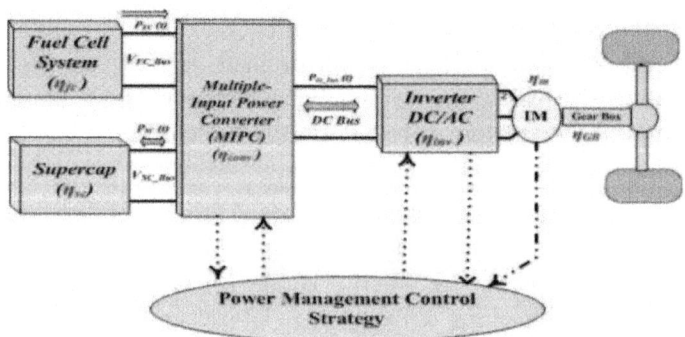

Figure 16: The drive system of the Fuel Cell/Supercapacitor Hybrid Electric Vehicle

Modeling of the vehicle power demand

The load force of the vehicle consists of gravitational force, rolling resistance, aerodynamic drag force, and acceleration force. Hereby, the load power required for vehicle acceleration can be written as [Hegazy & Van Mierlo, 2010; Hegazy et. al 2010]

$$P_{load} = \frac{(F_g + F_{roll} + F_{AD} + F_{acc}) * V}{\eta_{GB}} \tag{61}$$

$$F_g = M.g.\sin(\alpha) \tag{62}$$

$$F_{roll} = M.g.f_r.\cos(\alpha) \tag{63}$$

$$F_{AD} = 0.5 \rho_a . C_D . A_F . V^2 \tag{64}$$

$$F_{acc} = M. \frac{dV}{dt} \tag{65}$$

$$V = \omega_w . r_w \tag{66}$$

The total electric power required from sources can be expressed as:

$$P_{req} = \frac{P_{load}}{\eta_m . \eta_{Inv} . \eta_{Conv}} \tag{67}$$

The parameters of the vehicle are given in Table 6. The analysis of FCHEV is performed with two standard driving cycles:

1. The Federal Test Procedure (FTP75) Urban;

2. The New European Driving Cycle (NEDC)

Suppose that the efficiencies of the motor (η_m), inverter (η_{Inv}), and MIPEC ($\eta_{Conv} = \eta_B = \eta_{B/B}$) are 0.90, 0.94 and 0.95, respectively.

Table 6: Vehicle Parameters [Wu & Gao, 2006]

M	Vehicle mass (kg)	1450	A_f	Front Area (m2)	2.13
f_r	Rolling Resistance Coefficient	0.013	r_ω	Radius of the wheel (m)	0.28
C_D	Aerodynamic Drag Coefficient (CD)	0.29	ρ_a	Air density (kg/m3)	1.202

Optimal powertrain design

The first goal of optimization algorithm, PSO, is to minimize the cost, the mass, and the volume of the fuel cell (FC) and supercapacitor (SC). It is assumed that, the cost, the mass and the volume of the fuel cell and supercapacitor are a function of the number of the parallel units N_{fcp} and N_{scp}, respectively. The multi-objective criterion should be aggregated in a single objective function if the design objective is to embody a unique solution. The objective function can be formulated as follows:

$$F(x) = w_1 \cos t + w_2 \, mass + w_3 volume \tag{68}$$

$$\cos t = C1. \, Nfcs. \, Nfcp + C2. \, Nscs. \, Nscp \tag{69}$$

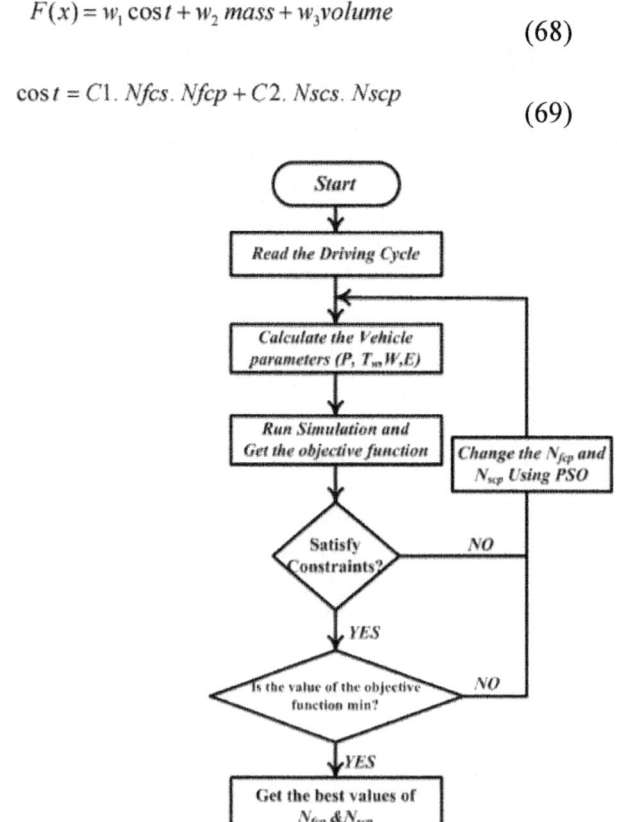

Figure 17: The flowchart of the execution of PSO [Hegazy et. al 2010]

The coefficients of the terms in F(x) were chosen to reflect the importance of minimizing the cost, the mass and the volume. Suppose that w1, w2, and w3 are 0.35, 0.35, and 0.3, respectively. Figure 17 presents the flowchart of the execution of PSO, which evaluates the optimal number of the FC units and the supercapacitor units by using MATLAB /SIMULINK. The layout of the fuel-cell stack and layout of the supercapacitor system are shown in Fig.18 (a) and (b), respectively. The constraints of the optimization problems are mentioned in [Hegazy & Van Mierlo, 2010].

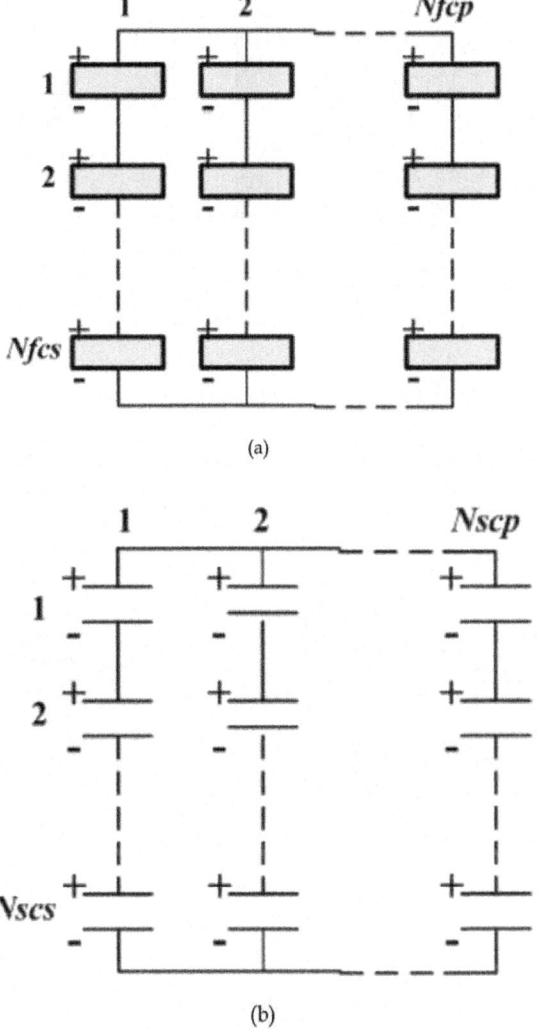

Figure 18: (a) Layout of the FC; (b) Layout of the SC

Based on minimizing the objective function F(x) in (68), the results of the optimal design and components sizing of the FC/SC powertrain are shown in Fig.19. The analyses and parameters of the FC and the SC are mentioned in [Hegazy & Van Mierlo, 2010].

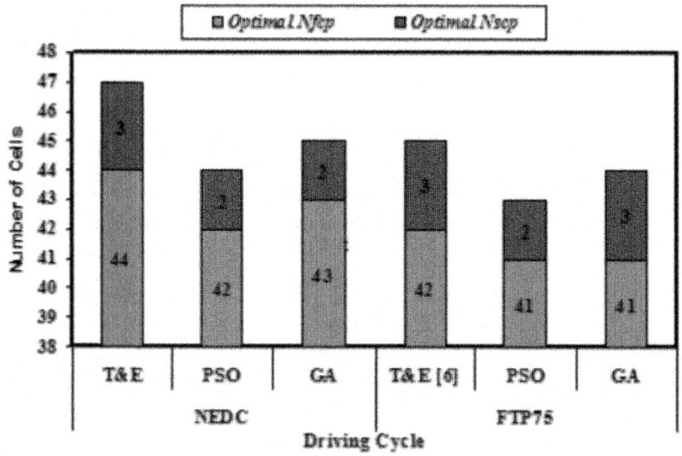

(a) The optimal numbers of cells of FC and SC

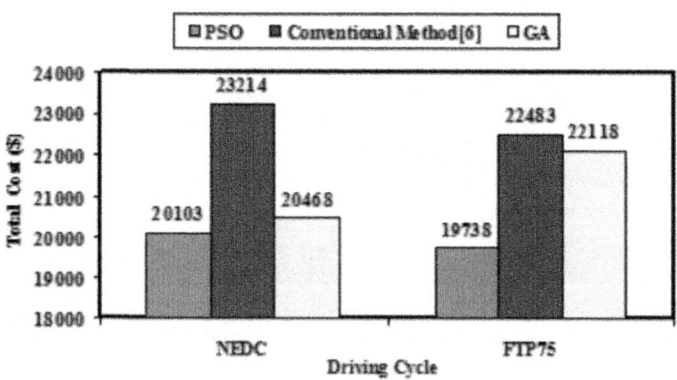

(b) The cost of the FC/SC components

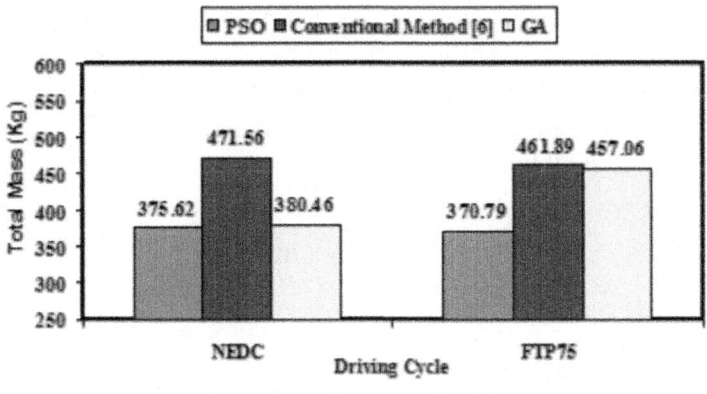

(c) The mass of the FC/SC components

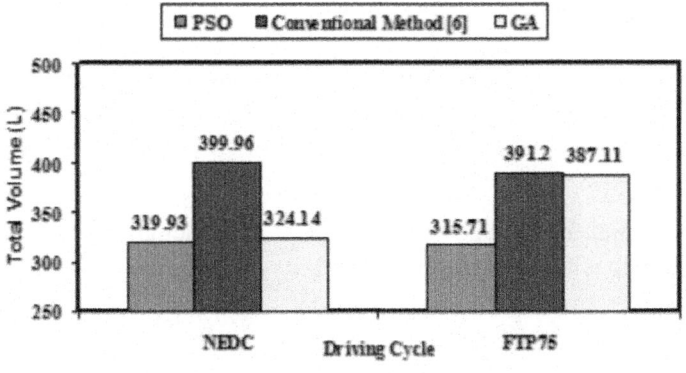

(d) The mass of the FC/SC components

Figure 19: The Comparative of the optimal design between different methods for FC/SC HEV

Optimal Power Control (OPC)

The second goal of the PSO is to minimize the vehicle fuel, hydrogen, consumption while maintaining the supercapacitor state of charge. As a hybrid powertrain is under consideration, a power management strategy is required to define what both the FC and SC powers are. The global optimization algorithms, such as GA and dynamic programming (DP), achieve an optimal power control for FC/SC hybrid electric vehicle, which leads to the lowest hydrogen consumption and maintains the supercapacitor SOC [Sinoquet et. al 2009; Sundstrom & Stefanopoulou 2006].

In this study, the optimal power control can be achieved by using PSO and GA for a given driving cycle. Suppose that the degree of hybridization of the fuel cell is K_{fc} at time t and K_{soc}, Proportional controller gain, which used to adapt the SOC during charging from the FC. A balance equation can naturally be established, since the sum of power from both sources has to be equal to the required power at all times:

$$P_{req}(t) = Pfc(t) + Psc(t) \tag{70}$$

$$Kfc(t) = \frac{Pfc(t)}{P_{req}(t)} \tag{71}$$

The net energy consumed from the FC at time t can be computed as follows:

$$Efc(t) = \int_0^t \frac{Pfc(t)}{\eta(Pfc(t))} dt \tag{72}$$

The cost function can be expressed as follows:

$$F_2(x) = \frac{1}{Elow} \sum_{K=0}^{N} \frac{Pfc_{Opti}(k)}{\eta(Pfc_{Opti}(k))} \Delta T \tag{73}$$

The Optimal fuel cell power output, P_{fcOpti}, is calculated based on the SOC of the supercapacitor and power demand, P_{req}, as follows:

$$Pfc_{Opti}(k) = Kfc(k) P_{req}(k) + Ksoc(k)\,(Pfc_{max} - Pfc_{min}) \left[\frac{SOC_{ref} - SOC(k)}{(SOC_{max} - SOC_{min})/2} \right] \tag{74}$$

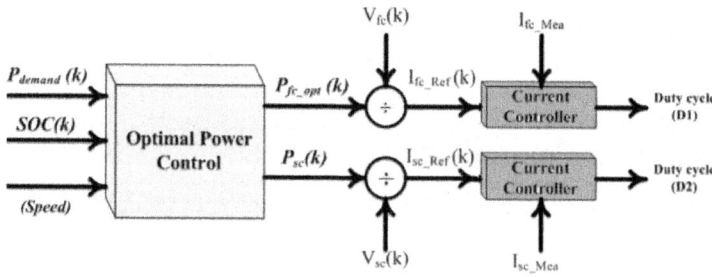

Figure 20: The block diagram of the Optimal power Control

Where: $N = T/T$ is number of samples during the driving cycle, and $T=1s$ is the sampling time. The block diagram of the optimal power control based on optimization algorithm is shown in Fig.20.

Based on minimizing the objective function $F_2(x)$ in (73), the results of the optimal power sharing based PSO and the comparative study for the FC/SC powertrain are summarized in Fig.21 [Hegazy et. al 2010]

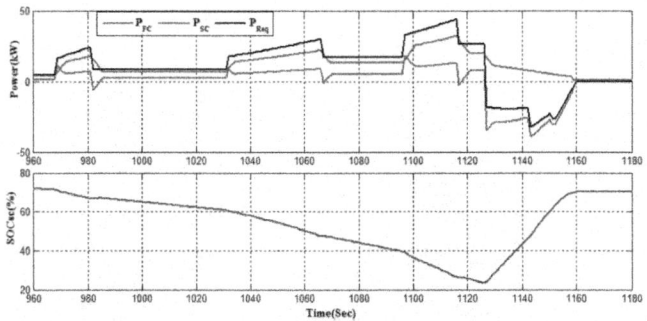

(a) The power sharing between FC and SC on NEDC driving cycle

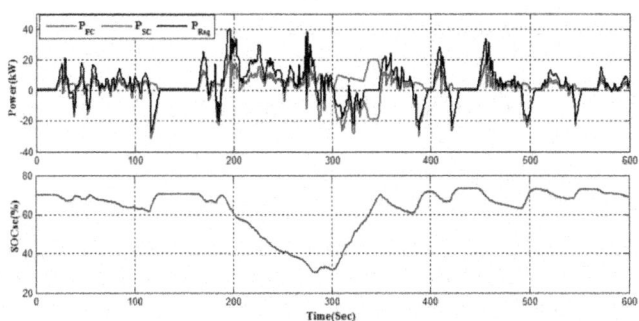

(b) The power sharing between FC and SC on FTP75 driving cycle

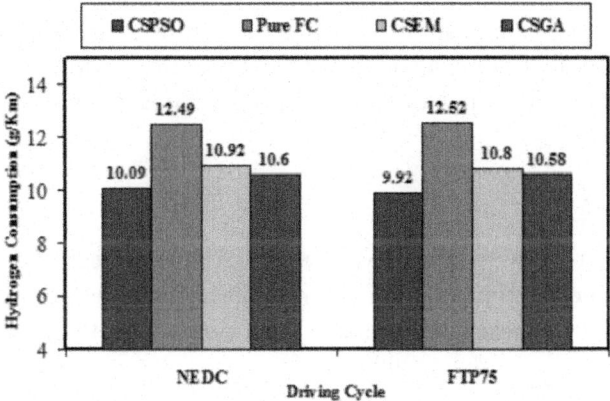

(c) The Comparative of the hydrogen consumption between control strategies

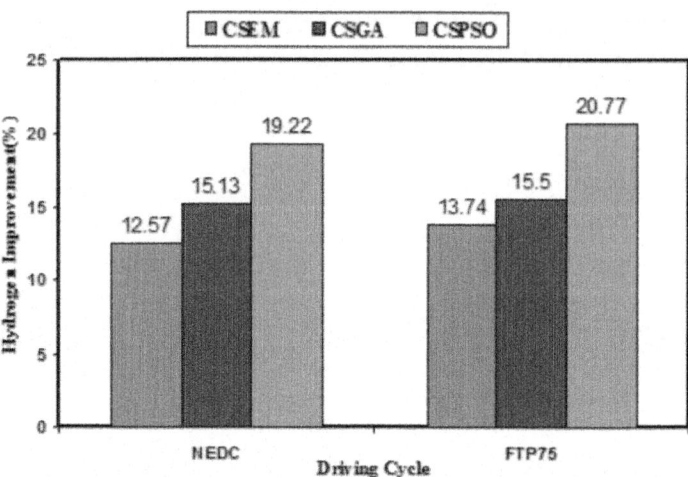

(d) The Hydrogen improvements with respect to pure fuel cell without SC

Figure 21: The results of the optimal power Control for FC/SC

CONCLUSION

This chapter deals with the applicability of swarm intelligence (SI) in the form of particles swarm optimization (PSO) used to achieve the best performance for the electric machines and electric drives. In addition, by analyzing and comparing the results, it is shown that control strategy based on PSO is more efficient than others control strategies to achieve the optimal performance for fuel cell/supercapacitor hybrid electric vehicles (FCHEV). It is very important to note that, these applications were achieved without any additional hardware cost, because the PSO is a software scheme. Consequently, PSO has positive promises for a wide range of variable speed drive and hybrid electric vehicles applications.

REFERENCES

1. Amin. A. M. A., Korfally. M. I., Sayed. A. A. and Hegazy. O.T. M., (2009), Efficiency Optimization of Two Asymmetrical Windings Induction Motor Based on Swarm Intelligence, IEEE Transactions on Energy Conversion, Vol. 24, No. 1, March 2009

2. Amin. A. M. A., Korfally. M. I., Sayed. A. A. and Hegazy. O.T. M., (2006), Losses Minimization of Two Asymmetrical Windings Induction Motor Based on Swarm Intelligence, Proceedings of IEEE- IECON 06 ,

pp 1150 – 1155, Paris, France, Nov. 2006.
3. Amin. A. M. A., Korfally. M. I., Sayed. A. A. and Hegazy. O.T. M., (2007), Swarm Intelligence-Based Controller of Two-Asymmetrical Windings Induction Motor, accepted for IEEE. EMDC07, pp 953 –958, Turkey, May 2007.
4. Eberhart. R, Kennedy. J, (1995), A New Optimizer Using Particles Swarm Theory, Proc.
5. Sixth International Symposium on Micro Machine and Human Science (Nagoya, Japan), IEEE Service Center, Piscataway, NJ, pp. 39-43,
6. Hamid Radwan H., Amin Amr. M. A., Ahmed Refaat S., and El-Gammal Adel A. A. ,(2006), New Technique For Maximum Efficiency And Minimum Operating Cost Of Induction Motors Based On Particle Swarm Optimization (PSO)", Proceedings of IEEE- IECON 06, pp 1029 – 1034, Paris, France, Nov. 2006.
7. Hegazy Omar, (2006), Losses Minimization of Two Asymmetrical Windings Induction Motor Based on Swarm Intelligence, M.Sc., Helwan University, 2006.
8. Hegazy Omar, and Van Mierlo Joeri, (2010), Particle Swarm Optimization for Optimal Powertrain Component Sizing and Design of Fuel cell Hybrid Electric Vehicle, 12th International Conference on Optimization of Electrical and Electronic Equipment, IEEE OPTIM 2010
9. Hegazy Omar, Van Mierlo Joeri, Verbrugge Bavo and Ellabban Omar, (2010), Optimal Power Sharing and Design Optimization for Fuel Cell/ Battery Hybrid Electric Vehicles Based on Swarm Intelligence, The 25th World Battery, Hybrid and Fuel Cell Electric Vehicle Symposium & Exhibition © EVS-25 Shenzhen, China, Nov. 5-9, 2010.
10. Kennedy. J and Eberhart .R, (2001), Swarm Intelligence, Morgan Kaufmann Publishers, Inc., San Francisco, CA
11. Kioskeridis, I; Margaris, N., (1996), Losses minimization in scalar-controlled induction motor drives with search controllers" Power Electronics, IEEE Transactions, Volume: 11, Issue: 2, March 1996 Pages: 213 – 220
12. Popescu. M, Navrapescu. V, (2000) ,A method of Iron Loss and Magnetizing Flux Saturation Modeling in Stationary Frame Reference of Single and Two –Phase Induction Machines", IEE 2000, Conf. power Elec. & Variable Speed Drives, 140-146
13. Sundstrom Olle and Stefanopoulou Anna, (2006), Optimal Power Split in Fuel Cell Hybrid Electric Vehicle with different Battery Sizes, Drive

Cycles, and Objectives, Proceedings of the 2006 IEEE International Conference on Control Applications Munich, Germany, October 4-6, 2006.
14. Van Mierlo Joeri, Cheng Yonghua, Timmermans Jean-Marc and Van den Bossche Peter, (2006), Comparison of Fuel Cell Hybrid Propulsion Topologies with Super-Capacitor, IEEE, EPE-PEMC 2006, Portorož, Slovenia
15. Wu Ying, Gao Hongwei, (2006) ,Optimization of Fuel Cell and Supercapacitor for Fuel-Cell Electric Vehicles, IEEE Transactions On Vehicular Technology, Vol. 55, No. 6, November 2006

Chapter 11

OPERATION OF ACTIVE FRONT-END RECTIFIER IN ELECTRIC DRIVE UNDER UNBALANCED VOLTAGE SUPPLY

Miroslav Chomat

Institute of Thermomechanics AS CR, v.v.i. Czech Republic

INTRODUCTION

Non-standard conditions in the power network such as voltage unbalance can negatively affect operation of electric drives. The unbalance can be caused by a failure in the network or by an unbalanced load in the electric vicinity of the affected drive. Unsymmetrical voltages at the input of a voltage source inverter cause pulsations in the DC link voltage when not properly taken care of. This may result in significantly reduced power capabilities and, therefore, limited controllability of the drive. This text deals with the effects of unbalanced voltage supply on the DC-link voltage pulsations, methods to address this problem and the additionally imposed constraints in operating regions of the rectifier.

CONTROL METHOD

A simplified scheme of the drive under investigation is shown in Fig. 1. The front-end controlled rectifier is connected to the mains through input filter inductors. The output current of the rectifier supplies the DC current to the output inverter and maintains the voltage across the DC-link capacitors constant at the same time. The value of this current can be controlled by suitable switching of solid-state elements in the front-end stage.

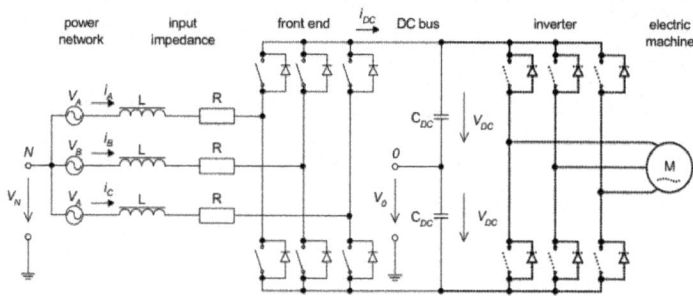

Figure 1: Scheme of system under investigation.

Suitable control of the front-end AC/DC converter can be employed in order to draw constant input power from the power network even at unbalanced voltage supply (Stankovic & Lipo, 2001; Lee et al., 2006; Cross et al., 1999; Song & Nam, 1999). The switching functions for the front-end AC/DC converter are generated so that a constant voltage across the DC bus is maintained. Series combinations of inductance and resistance are considered at the input terminals of the inverter.

The system can be electrically described by the following set of ordinary differential equations (Chomat & Schreier, 2005):

$$v_A - L\frac{di_A}{dt} - Ri_A - v_{SA} + v_N - v_0 = 0, \tag{1}$$

$$v_B - L\frac{di_B}{dt} - Ri_B - v_{SB} + v_N - v_0 = 0, \tag{2}$$

$$v_C - L\frac{di_C}{dt} - Ri_C - v_{SC} + v_N - v_0 = 0, \tag{3}$$

where

$$v_{SA} = s_A \cdot V_{DC}, \tag{4}$$

$$v_{SB} = s_B \cdot V_{DC}, \tag{5}$$

$$v_{SC} = s_C \cdot V_{DC} \tag{6}$$

are the voltages at the input of the inverter. The functions s_A, s_B, and s_C are the corresponding unit switching functions of the particular phases of the front-end stage, which represent the fundamental harmonic components of the pulse width modulated output. Sinusoidal switching functions with the nominal frequency are considered throughout this paper, whereas the higher harmonics that would arise in a real power converter are neglected in the calculation for simplification. V_{DC} represents one half of the overall DC-link voltage here. The voltage v_N is the electric potential of the neutral of the mains and v_0 is the electric potential of the centre point of the capacitor bank in the DC bus. The DC-link current can be calculated from the phase currents and the switching functions according to

$$i_{DC} = \frac{1}{2}(s_A i_A + s_B i_B + s_C i_C). \qquad (7)$$

The coefficient ½ takes into account the fact that currents in both positive and negative directions that flow through different current paths in the DC bus are produced by the rectifier.

An unbalanced system of phase quantities can advantageously be represented by phasors of positive and negative rotating sequences. It is not necessary to take zero-sequence quantities into account here as no neutral wire is considered in the system and, therefore, no zero-sequence current can develop. The resulting rotating vector of such a quantity may then be written as

$$\mathbf{x} = \mathbf{X_P} e^{j\omega t} + \mathbf{X_N} e^{-j\omega t}. \qquad (8)$$

The subscripts P and N denote the positive and negative rotating sequences, respectively. Based on these assumptions, (1) - (3) and (4) – (6) may be rewritten in phasor form as

$$\mathbf{V_P} - (R + j\omega L)\mathbf{I_P} - \mathbf{S_P} V_{DC} = 0, \qquad (9)$$

$$\mathbf{V_N} - (R - j\omega L)\mathbf{I_N} - \mathbf{S_N} V_{DC} = 0. \qquad (10)$$

The solution of (9) and (10) for positive and negative sequence currents is

$$\mathbf{I_P} = \frac{\mathbf{V_P} - \mathbf{S_P} V_{DC}}{R + j\omega L}, \qquad (11)$$

$$\mathbf{I_N} = \frac{\mathbf{V_N} - \mathbf{S_N} V_{DC}}{R - j\omega L} \qquad (12)$$

and the corresponding rotating vector of the input currents is therefore

$$\mathbf{i} = \mathbf{I_P} e^{j\omega t} + \mathbf{I_N} e^{-j\omega t}. \qquad (13)$$

Similarly, we can formally introduce a rotating vector of the switching functions

$$s = S_P e^{j\omega t} + S_N e^{-j\omega t}. \qquad (14)$$

Then the resulting instantaneous value of the current supplied into the DC link by the front-end converter from (7) can be written in the vector form as

$$i_{DC} = \frac{1}{2}\frac{3}{2}\text{Re}\{\mathbf{i}\cdot\bar{\mathbf{s}}\}, \qquad (15)$$

where the bar over the symbol denotes the complex conjugate value. The term 3/2 appears in (15) due to the transformation from rotating vector form to instantaneous quantities.

The resulting relation obtained after substituting (13) and (14) into (15) can be written as the sum of two separate current components and given as

$$i_{DC} = i_{DC(avg)} + i_{DC(2\omega t)}, \qquad (16)$$

where

$$i_{DC(avg)} = \frac{3}{4}\text{Re}\left\{\frac{V_P - S_P V_{DC}}{R + j\omega L}\bar{S}_P + \frac{V_N - S_N V_{DC}}{R - j\omega L}\bar{S}_N\right\}, \qquad (17)$$

$$i_{DC(2\omega t)} = \frac{3}{4}\text{Re}\left\{\frac{V_P - S_P V_{DC}}{R + j\omega L}\bar{S}_N e^{j2\omega t} + \frac{V_N - S_N V_{DC}}{R - j\omega L}\bar{S}_P e^{-j2\omega t}\right\}. \qquad (18)$$

The first component, (17), represents a DC component and the second, (18), represents a pulsating component with the frequency twice as high as that of the mains. The pulsating component is only produced when the negative sequence of either the input voltages or the switching functions is present.

From (18), a condition for the elimination of the pulsating component in the DC link can be derived

$$\text{Re}\left\{\frac{V_P - S_P V_{DC}}{R + j\omega L}\bar{S}_N e^{j2\omega t}\right\} = -\text{Re}\left\{\frac{V_N - S_N V_{DC}}{R - j\omega L}\bar{S}_P e^{-j2\omega t}\right\}. \qquad (19)$$

As the real part of a complex number equals the real part of its conjugate value, (18) can also be written as

$$\text{Re}\left\{\frac{V_P - S_P V_{DC}}{R + j\omega L}\bar{S}_N e^{j2\omega t}\right\} = -\text{Re}\left\{\frac{\bar{V}_N - \bar{S}_N V_{DC}}{R + j\omega L} S_P e^{j2\omega t}\right\}. \qquad (20)$$

For the above equation to be satisfied at any time, the following must hold providing that there is non-zero input impedance

$$(\mathbf{V_P} - \mathbf{S_P} V_{DC})\bar{\mathbf{S}}_\mathbf{N} = -(\bar{\mathbf{V}}_\mathbf{N} - \bar{\mathbf{S}}_\mathbf{N} V_{DC})\mathbf{S_P}. \qquad (21)$$

If the input voltages are known and the control is free to choose the positive sequence component of the switching functions, the negative sequence of the switching functions obtained from (21) is

$$\mathbf{S_N} = \frac{\bar{\mathbf{S}}_\mathbf{P} \mathbf{V_N}}{2\bar{\mathbf{S}}_\mathbf{P} V_{DC} - \bar{\mathbf{V}}_\mathbf{P}}. \qquad (22)$$

It should be noted that the relation does not contain values of input resistance and inductance and is, therefore, the same for pure inductance as well as for pure resistance connected to the front end of the inverter.

As the amplitudes of the individual switching functions need to be less than or equal to one, the range of practical combinations of $\mathbf{S_P}$ and $\mathbf{S_N}$ is constrained. A simple, and rather conservative, condition to keep the switching functions in allowable limits can be written as

$$|\mathbf{S_P}| + |\mathbf{S_N}| \le 1. \qquad (23)$$

For more precise evaluation of the constraints, we need to evaluate magnitudes of switching vectors in individual phases

$$|\mathbf{S}_A| = |\mathbf{S_P} + \bar{\mathbf{S}}_\mathbf{N}|, \qquad (24)$$

$$|\mathbf{S}_B| = |\mathbf{S_P} a + \bar{\mathbf{S}}_\mathbf{N} a^2|, \qquad (25)$$

$$|\mathbf{S}_C| = |\mathbf{S_P} a^2 + \bar{\mathbf{S}}_\mathbf{N} a| \qquad (26)$$

and limit the magnitude of each of them

$$(|\mathbf{S}_A| \le 1) \wedge (|\mathbf{S}_B| \le 1) \wedge (|\mathbf{S}_C| \le 1). \qquad (27)$$

For its operation, the above discussed control method requires to monitor the instantaneous values of the input phase voltages and of the DC-link voltage. Based on this information, a convenient combination of values of $\mathbf{S_P}$ and $\mathbf{S_N}$ can be chosen to produce the required value of the DC-link current and to satisfy the conditions in (22) and (23) at the same time. From $\mathbf{S_P}$ and $\mathbf{S_N}$, the switching functions for the individual legs of the rectifier are computed and switching pulses are generated for individual switching devices based on a particular pulse width modulation algorithm. The switching devices are considered to be transistors or thyristors with forced commutation.

OPERATION OF DRIVE UNDER UNBALANCED VOLTAGE SUPPLY

Numerical simulation of drive under unbalanced voltage supply

Operation of the described system has been numerically simulated under various types of unbalanced voltage supply in order to investigate the effect of the unbalance on the system behavior and the influence of certain circuit parameters. The reference parameters of the input impedance were chosen to be R = 0.1 Ω and L = 10 mH. The input phase voltages had nominal voltage amplitudes of 230 V, nominal frequency of 50 Hz, and mutual phase shifts of 120° to form a three-phase voltage system in the case of the symmetrical system. The DC-link voltage was set to 560 V and the capacitor of 1000 µF was used in the DC bus (Chomat et al., 2007).

First, operation of the investigated system under symmetrical voltage supply was simulated to obtain the reference case to compare with unbalanced operation. Figure 2 shows input phase voltages and currents and Figure 3 shows the DC-link current and voltage. It can be seen that both electrical quantities in the DC bus are smooth with no visible pulsations.

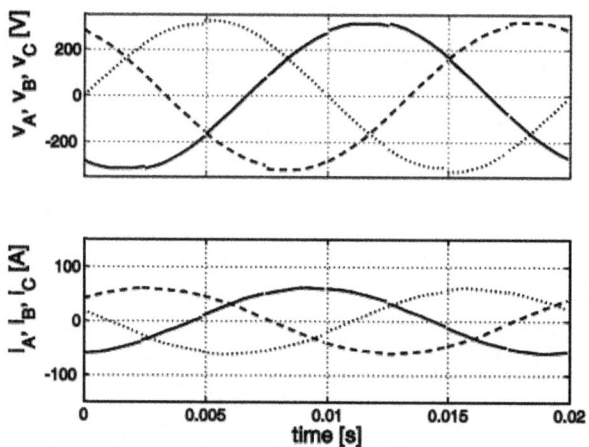

Figure 2: Phase voltages and currents under symmetrical voltage supply.

Second, the unbalance caused by setting the magnitude of the voltage in phase A to 200 V_{RMS} was investigated. Figures 4 and 5 show the corresponding quantities at the input of the rectifier and in the DC bus when no measures are taken to eliminate the pulsations by suitable modification of switching in the active front-end rectifier. The DC-link current and voltage contain significant pulsations that would make control of the drive more complicated. When the switching functions are modified in order to eliminate the effect of the supply voltage unbalance, the pulsations are nearly entirely eliminated, Figures 6 and 7. This has also an effect on the input phase currents compared to the previous case.

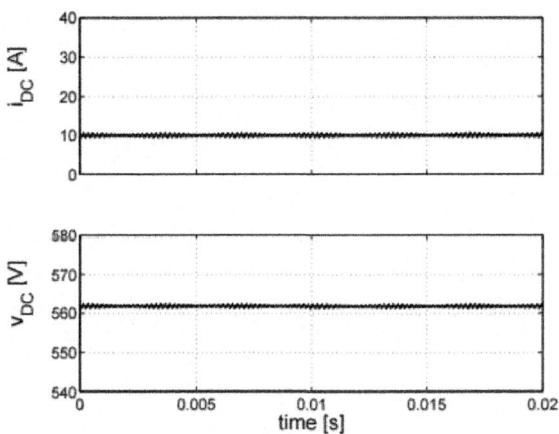

Figure 3: DC-link voltage and current under symmetrical voltage supply.

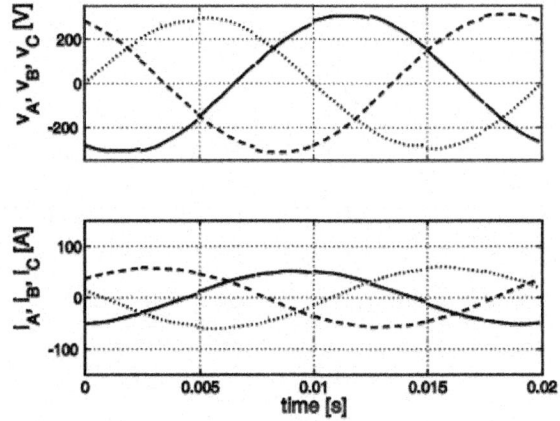

Figure 4: Phase voltages and currents under unbalanced voltage supply without compensation.

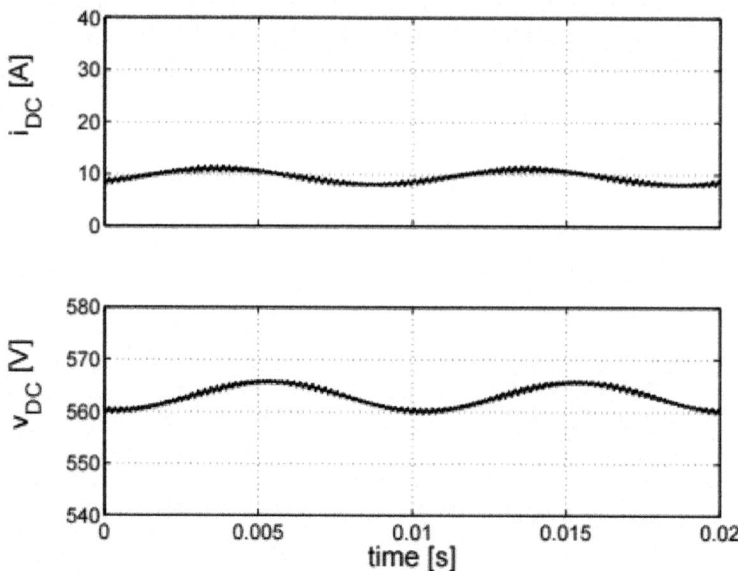

Figure 5: DC-link voltage and current under unbalanced voltage supply without compensation.

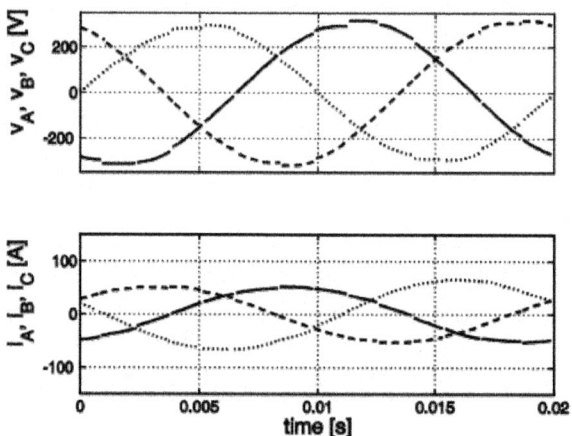

Figure 6: Phase voltages and currents under unbalanced voltage supply with compensation.

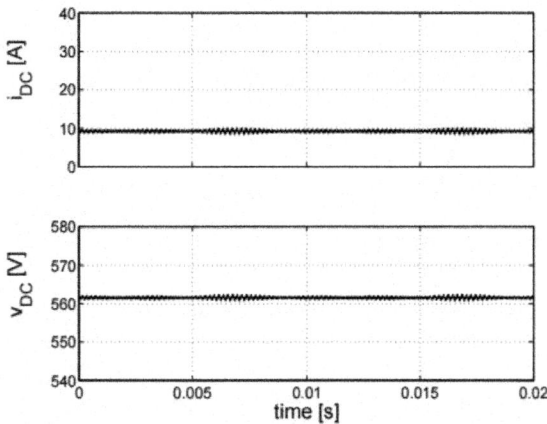

Figure 7: DC-link voltage and current under unbalanced voltage supply with compensation.

When DC-link capacitor is reduced to have the capacity of only 500 µF instead of 1000 µF, Figures 8 and 9, the DC-link voltage pulsations are increased twice as could be expected in case with no measures to eliminate the effect of the unbalance in the front-end rectifier. No change appears in the case when switching functions are modified to eliminate the effect of the unbalance, Figures 10 and 11.

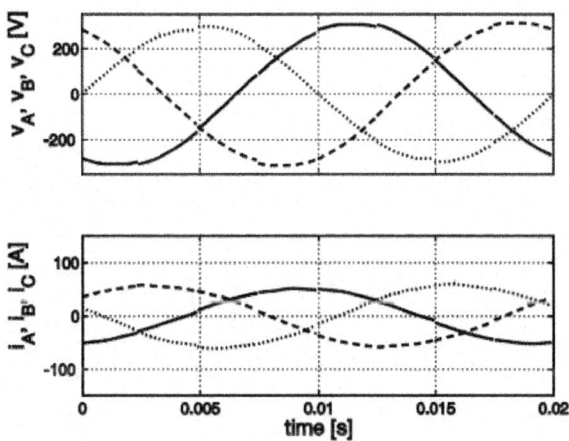

Figure 8: Phase voltages and currents under unbalanced voltage supply with reduced DC-link capacitance without compensation.

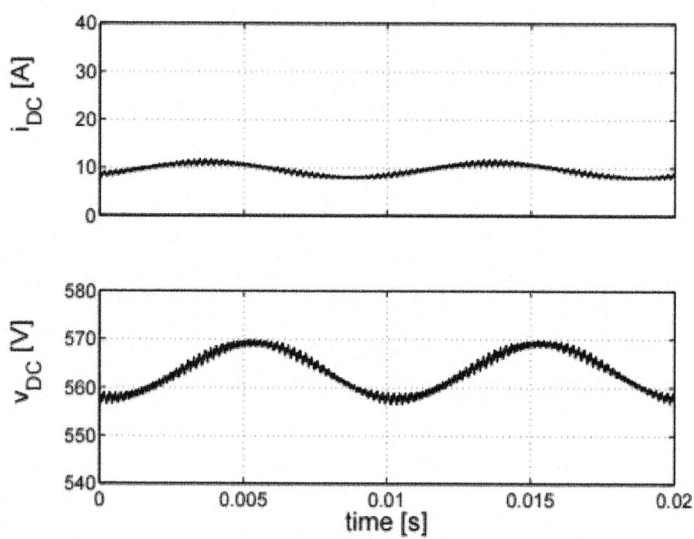

Figure 9: DC-link voltage and current under unbalanced voltage supply with reduced DC-link capacitance without compensation.

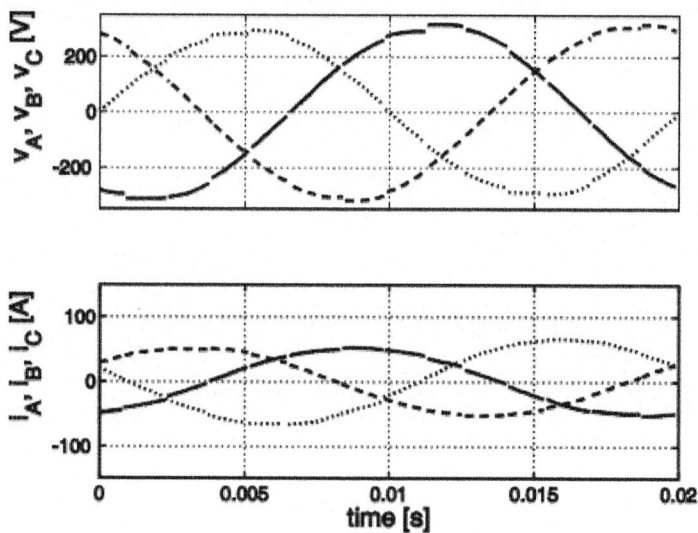

Figure 10: Phase voltages and currents under unbalanced voltage supply with reduced DC-link capacitance with compensation.

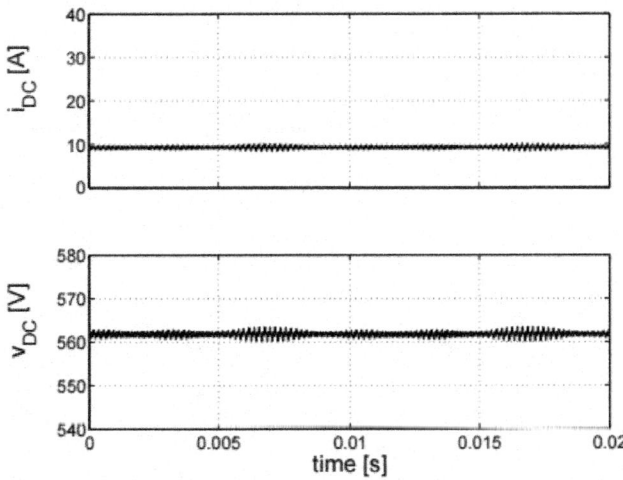

Figure 11: DC-link voltage and current under unbalanced voltage supply with reduced DC-link capacitance with compensation.

The change of the input inductance from 10 mH to 5 mH leads to an adequate increase in the input phase currents as well as in the DC-link current, Figures 12 to 15. The relative amount of pulsations remain at about the same levels.

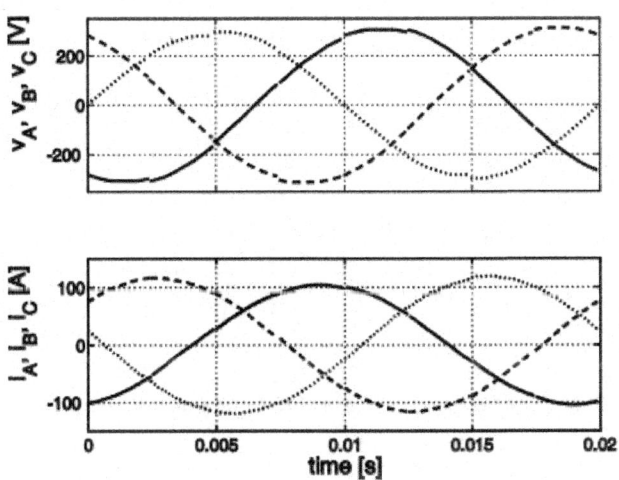

Figure 12: Phase voltages and currents under unbalanced voltage supply with reduced input inductance without compensation.

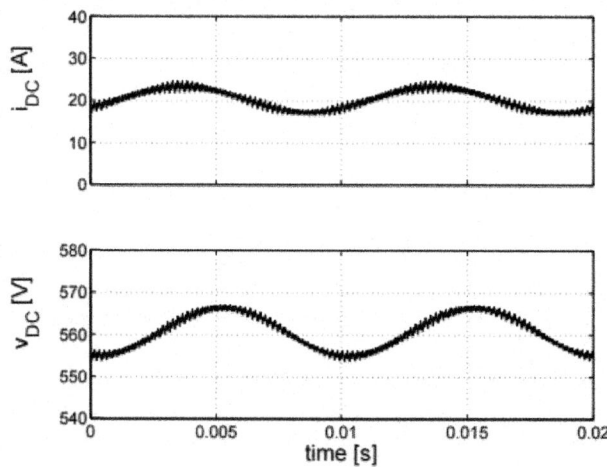

Figure 13: DC-link voltage and current under unbalanced voltage supply with reduced input inductance without compensation.

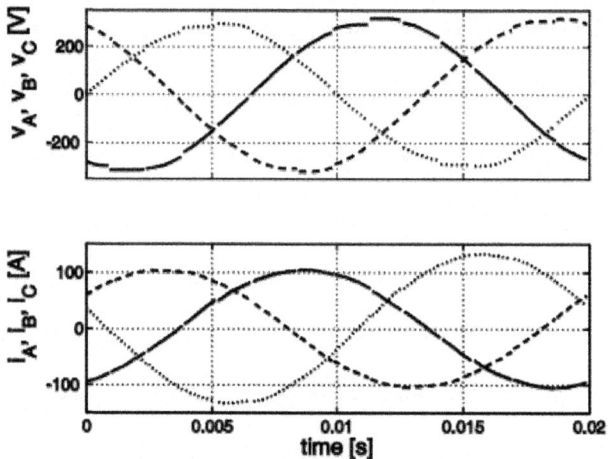

Figure 14: Phase voltages and currents under unbalanced voltage supply with reduced input inductance with compensation.

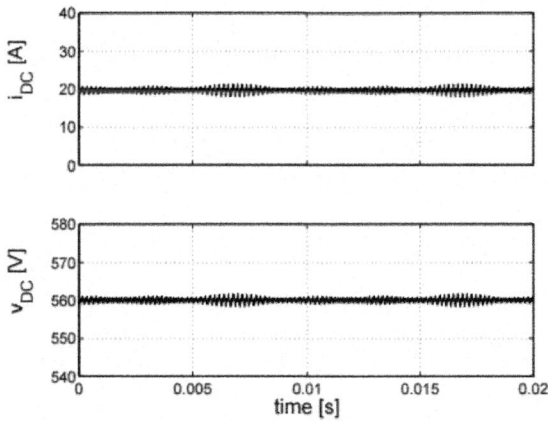

Figure 15: DC-link voltage and current under unbalanced voltage supply with reduced input inductance with compensation.

Finally, the unbalance caused by shifting the voltage phasor of phase *A* by 10° was investigated. Corresponding results due to the changes in circuit parameters are illustrated in Figures 16 to 27. It can be noted that the effects are similar to the previous case of unbalance.

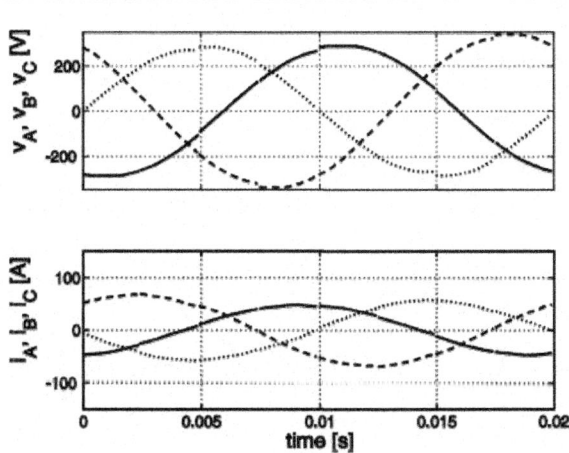

Figure 16: Phase voltages and currents under unbalanced voltage supply without compensation.

252   Electrical Machine Principles: A Handbook

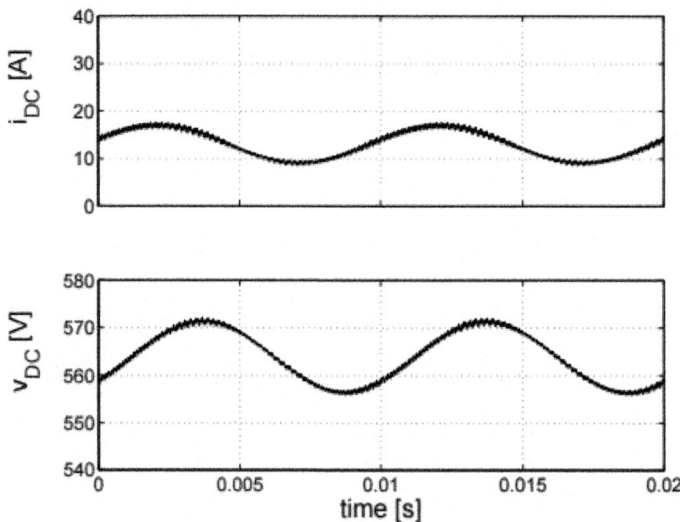

Figure 17: DC-link voltage and current under unbalanced voltage supply without compensation.

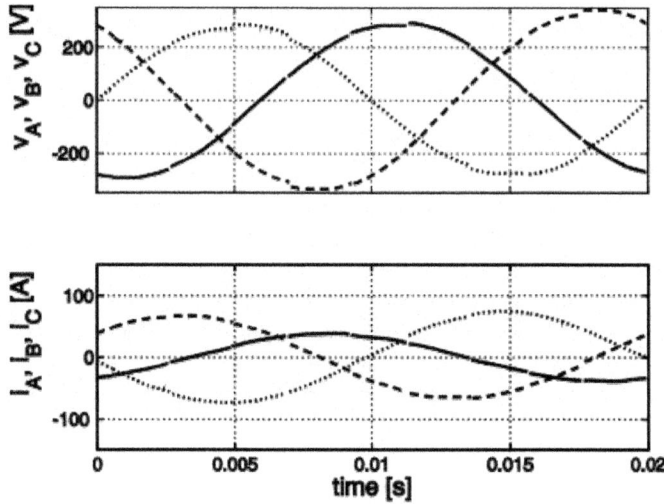

Figure 18: Phase voltages and currents under unbalanced voltage supply with compensation.

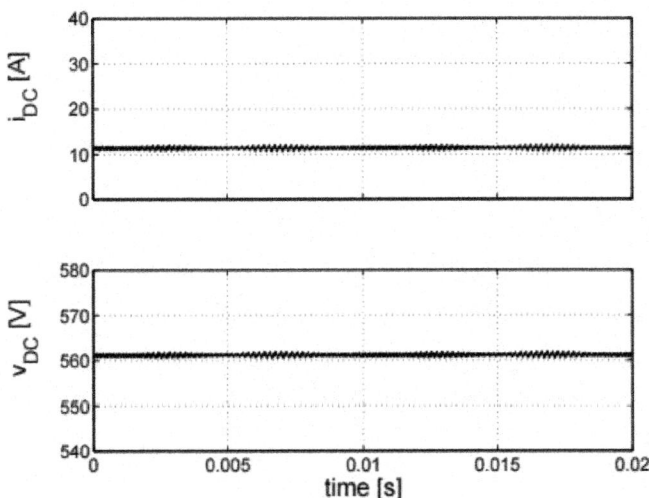

Figure 19: DC-link voltage and current under unbalanced voltage supply with compensation.

The effect of reduction of the DC-link capacitor from 1000 μF to 500 μF is shown in Figures 20 through 23.

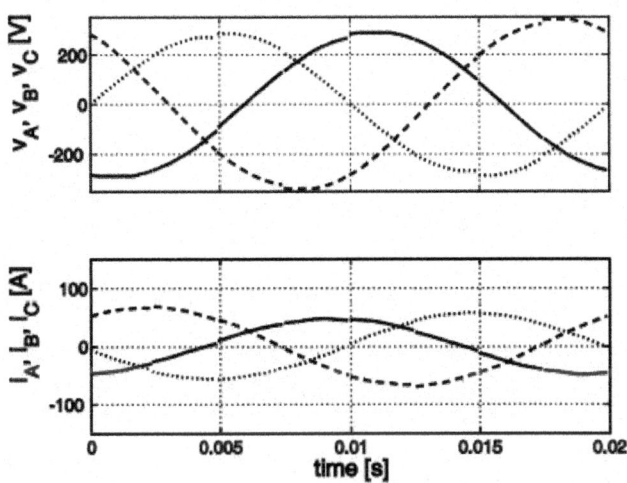

Figure 20: Phase voltages and currents under unbalanced voltage supply with reduced DC-link capacitance without compensation

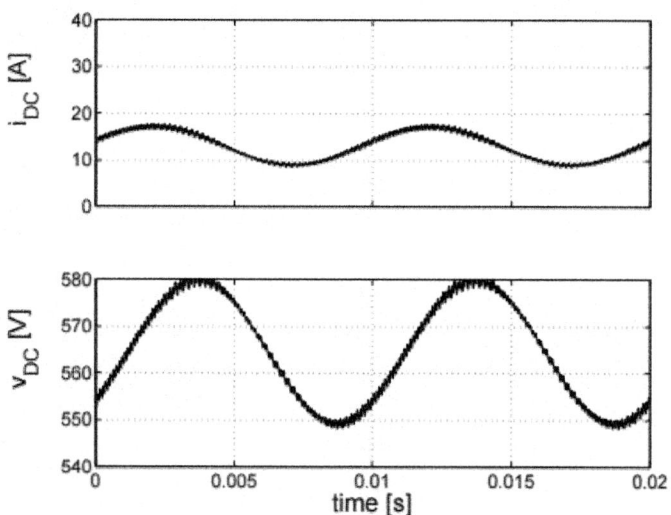

Figure 21: DC-link voltage and current under unbalanced voltage supply with reduced DC-link capacitance without compensation.

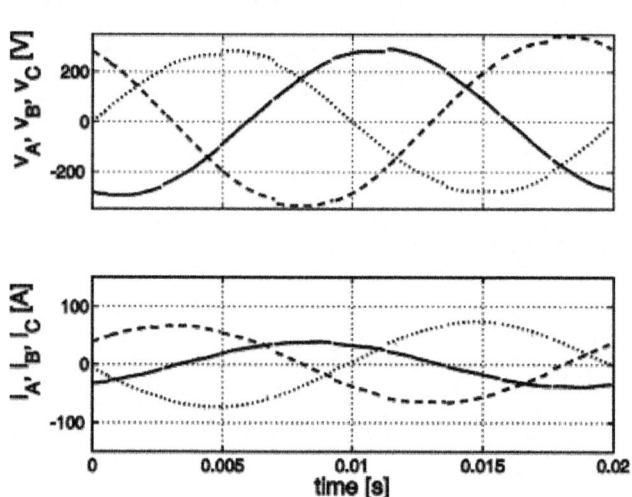

Figure 22: Phase voltages and currents under unbalanced voltage supply with reduced DC-link capacitance with compensation.

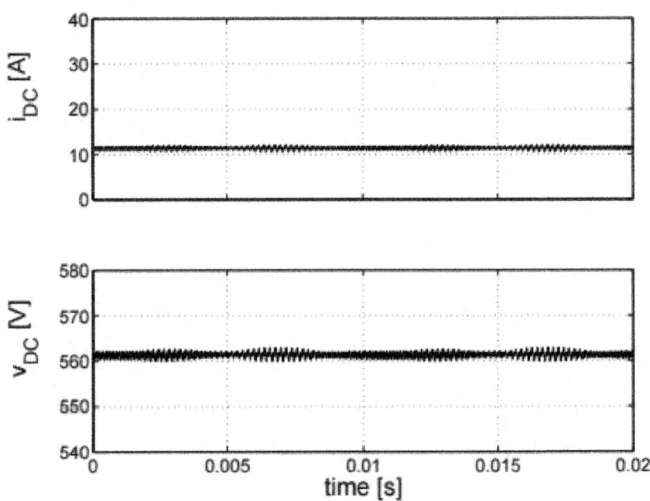

Figure 23: DC-link voltage and current under unbalanced voltage supply with reduced DC-link capacitance with compensation.

The corresponding situation for reduced input inductance from 10 mH to 5 mH is illustrated in Figures 24 to 27.

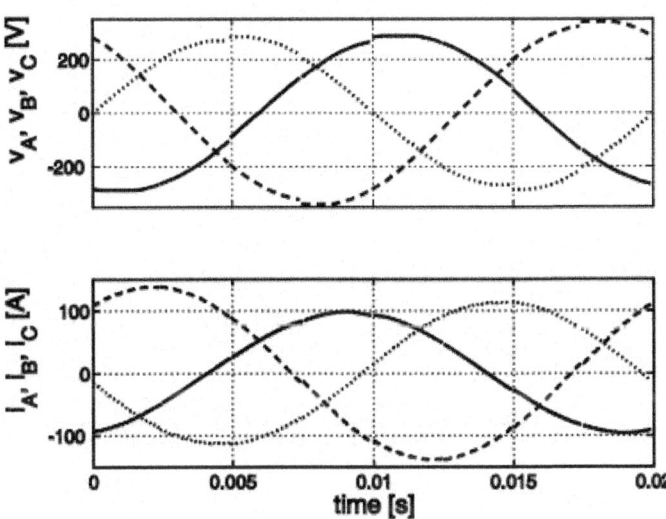

Figure 24: Phase voltages and currents under unbalanced voltage supply with reduced input inductance without compensation.

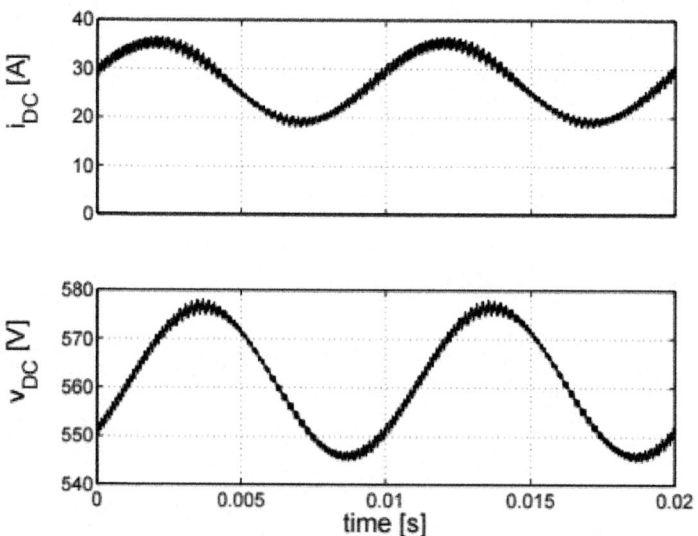

Figure 25: DC-link voltage and current under unbalanced voltage supply with reduced input inductance without compensation.

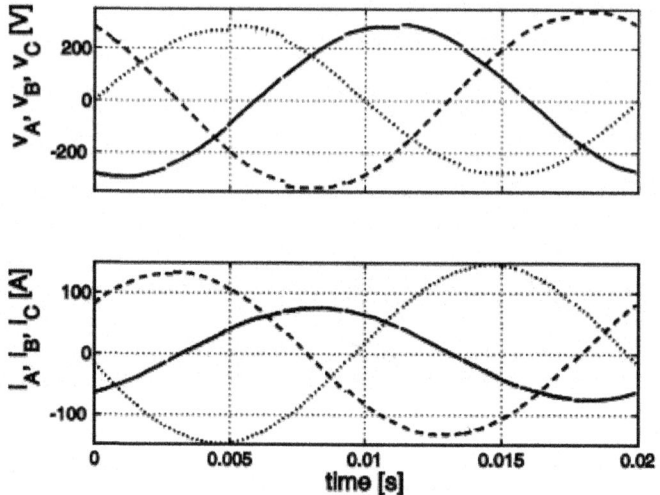

Figure 26: Phase voltages and currents under unbalanced voltage supply with reduced input inductance with compensation.

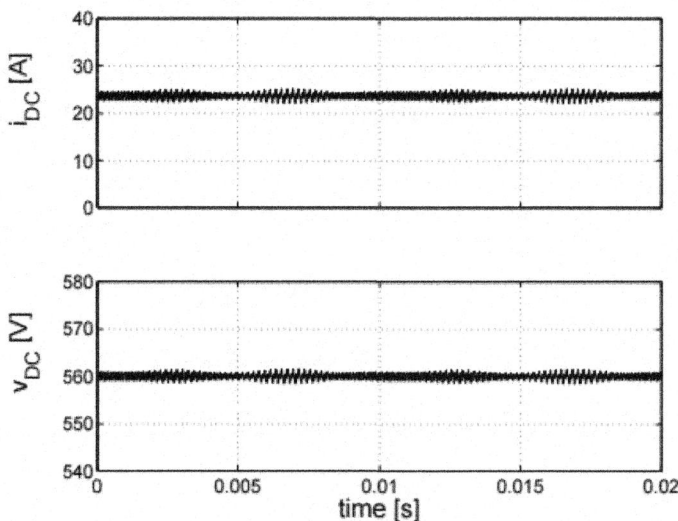

Figure 27: DC-link voltage and current under unbalanced voltage supply with reduced input inductance with compensation.

Limitation of control range due to unbalanced voltage supply

The necessity to generate the negative sequence component of the switching functions in order to eliminate the effect of the supply-voltage unbalance on the DC-link voltage pulsations reduces the control range for the positive sequence component of the switching functions (Chomat et al., 2009). This is due to the fact that in individual phases the maximum of the switching function can only reach one at most at any given time. Another constraint results from the current rating of the converter. The resulting constraints depend on the value and type of the unbalance.

Analysis of the limitation corresponding to various types of unbalanced supply voltages has been carried out. The reference parameters of the input impedance were chosen to be $R = 0.1$ and $L = 10$ mH. The input phase voltages had nominal voltage amplitudes of 230 V, nominal frequency of 50 Hz, and mutual phase shifts of 120° to form a three-phase voltage system in the case of the symmetrical system. The DC-link voltage was set to 400 V. The choice of the positive sequence component of the switching functions from the available control range affects both the magnitude of the DC-link current and the currents in individual input phases. Figure 28 shows what magnitudes of the DC-link current correspond to the coordinates from the available control range.

The unbalance was formed by setting the magnitude of the voltage in phase A to 0.75 p.u. The corresponding maximal input phase current magnitude, calculated as the maximum of all the phase currents, is shown in Figure 29. It can be seen from Figure 28 that the resulting DC-link current decreases in the vertical direction of the operating region, whereas the maximal input current in Figure 29 decreases in the horizontal direction. The corresponding measure of the current unbalance is depicted in Figure 30 and the average power factor of all the three input phases is depicted in Figure 31.

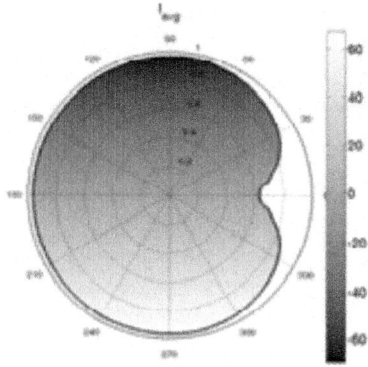

Figure 28: DC-link current under unbalanced voltage supply ($L = 10$ mH, $R = 0.1$ Ω, $V_{dc} = 400$ V).

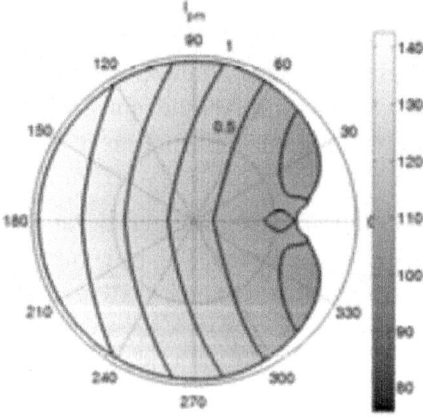

Figure 29: Maximal input phase current under unbalanced voltage supply ($L = 10$ mH, $R = 0.1$ Ω, $V_{dc} = 400$ V).

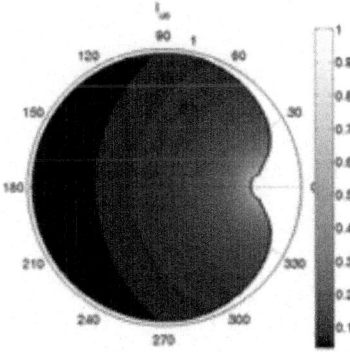

Figure 30: Input current unbalance under unbalanced voltage supply (L = 10 mH, R = 0.1 Ω, V_{dc} = 400 V).

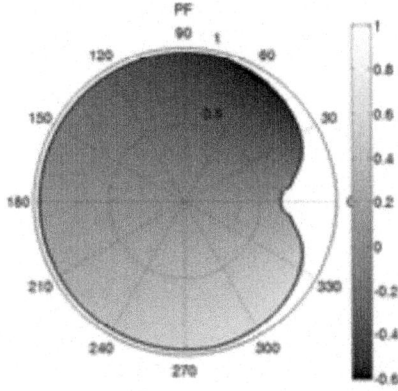

Figure 31: Power factor under unbalanced voltage supply (L = 10 mH, R = 0.1 Ω, V_{dc} = 400 V).

If we change the value of the input inductance from 10 mH to 1 mH, the constraints caused by the switching functions remain the same as can be seen from Figures 32 through 35. However, both the DC-link current and the input current increased nearly ten times as the input reactance represents the main limiting factor for the currents entering the rectifier. The excessive values of the currents would, in a case of a real rectifier, impose additional restrictions to the operating regions resulting from current stress of electronic components in the bridge. This can also be considered in the shape of new borders of operating regions.

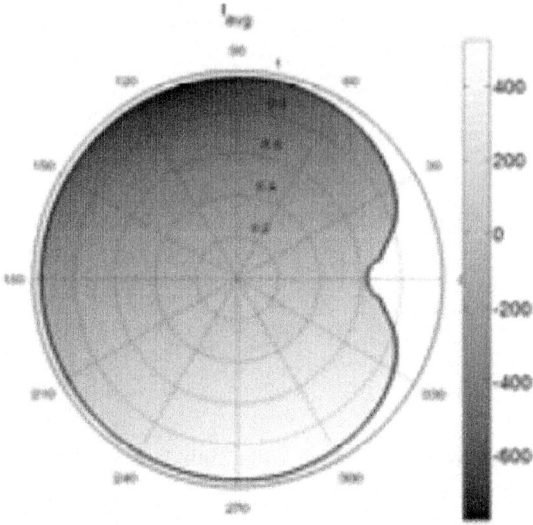

Figure 32: DC-link current under unbalanced voltage supply ($L = 1$ mH, $R = 0.1$ Ω, $V_{dc} = 400$ V).

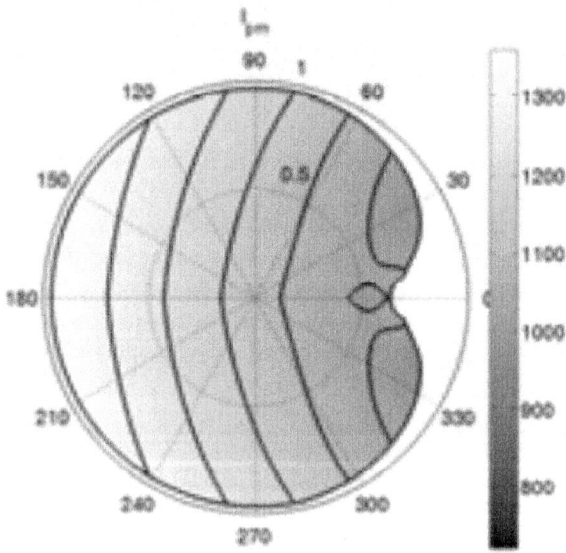

Figure 33: Maximal input phase current under unbalanced voltage supply ($L = 1$ mH, $R = 0.1$ Ω, $V_{dc} = 400$ V).

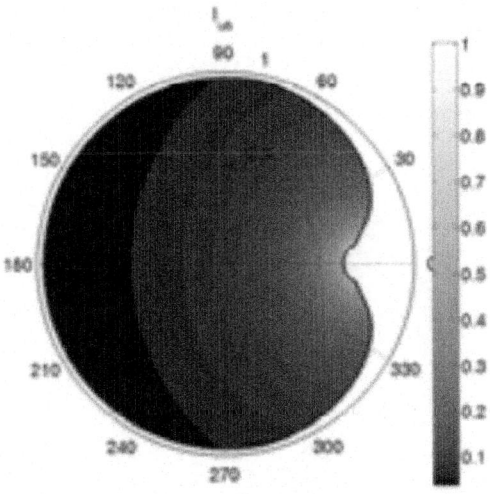

Figure 34: Input current unbalance under unbalanced voltage supply ($L = 1$ mH, $R = 0.1$ Ω, $V_{dc} = 400$ V).

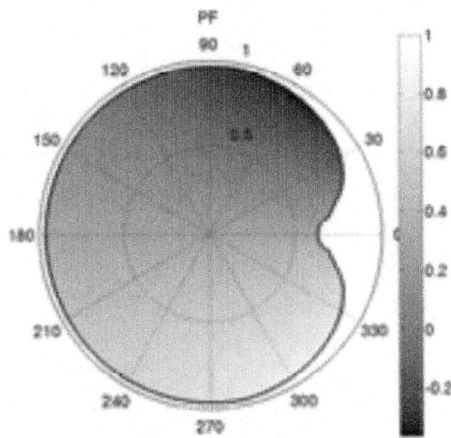

Figure 35: Power factor under unbalanced voltage supply ($L = 1$ mH, $R = 0.1$ Ω, $V_{dc} = 400$ V).

A different situation arises when the input resistance is increased ten times to 1 Ω. The corresponding electrical quantities are shown in Figures 36 through 39. The increase in the DC-link and input phase currents is not as dramatic as the resistance plays less significant role in limiting the currents than the inductance. The values of the currents are similar to the ones in the first case.

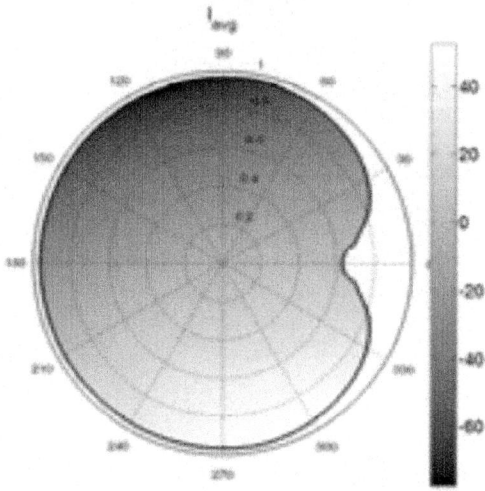

Figure 36: DC-link current under unbalanced voltage supply ($L = 1$ mH, $R = 1\ \Omega$, $V_{dc} = 400$ V).

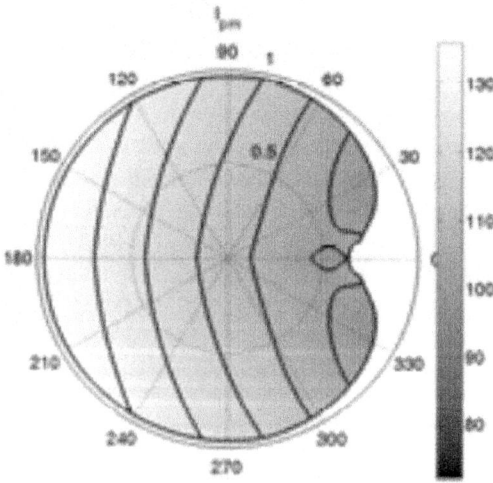

Figure 37: Maximal input phase current under unbalanced voltage supply ($L = 1$ mH, $R = 1\ \Omega$, $V_{dc} = 400$ V).

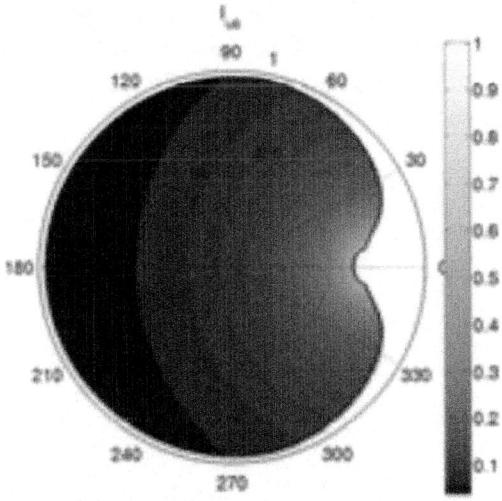

Figure 38: Input current unbalance under unbalanced voltage supply ($L = 1$ mH, $R = 1$ Ω, $V_{dc} = 400$ V).

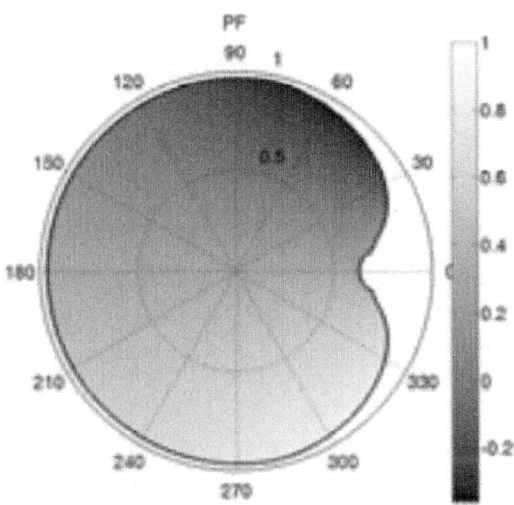

Figure 39: Power factor under unbalanced voltage supply ($L = 1$ mH, $R = 1$ Ω, $V_{dc} = 400$ V).

A change in the DC-link voltage introduces, on the other hand, a noticeable change in the shape of constraints caused by the limitation of the switching functions. Figures 40 through 43 show the situation for the decrease in the DC-link voltage from 400 V to 200 V and Figures 45 through 47 show the situation for the increase to 600 V. In the latter case, a rise of an isolated restricted area in the right hand side of the figure completely surrounded by available control space can be noticed.

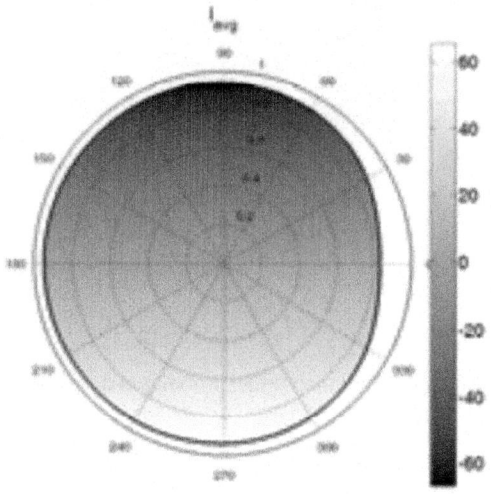

Figure 40: DC-link current under unbalanced voltage supply ($L = 10$ mH, $R = 0.1$ Ω, $V_{dc} = 200$ V).

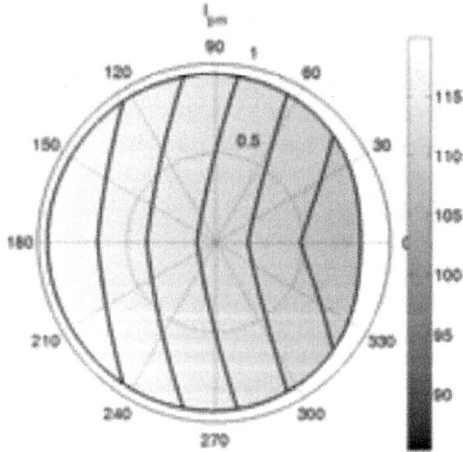

Figure 41: Maximal input phase current under unbalanced voltage supply ($L = 10$ mH, $R = 0.1$ Ω, $V_{dc} = 200$ V).

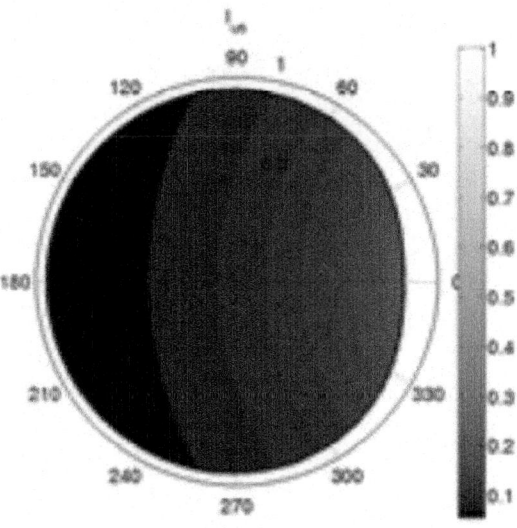

Figure 42: Input current unbalance under unbalanced voltage supply (L = 10 mH, R = 0.1 Ω, V_{dc} = 200 V).

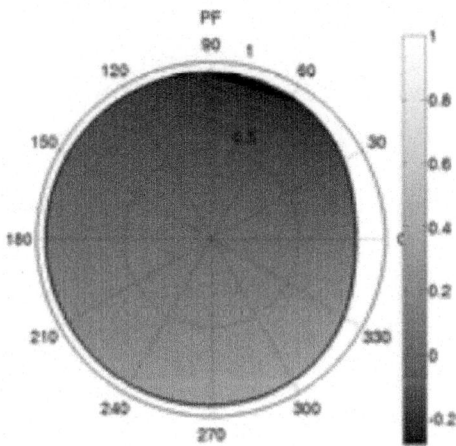

Figure 43: Power factor under unbalanced voltage supply (L = 10 mH, R = 0.1 Ω, V_{dc} = 200 V).

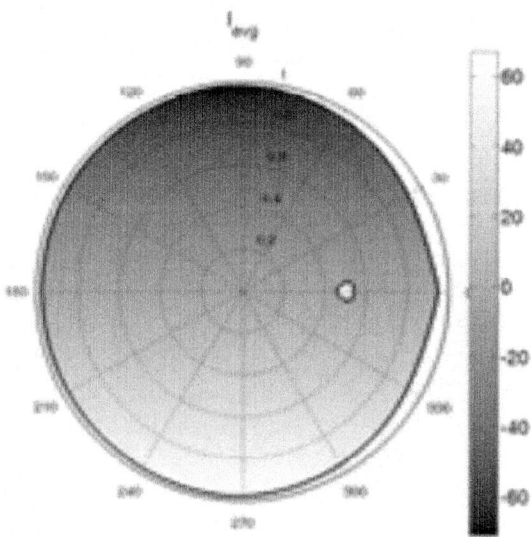

Figure 44: DC-link current under unbalanced voltage supply ($L = 10$ mH, $R = 0.1$ Ω, $V_{dc} = 600$ V).

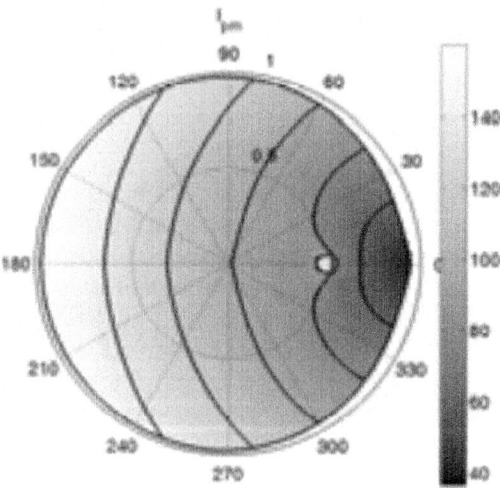

Figure 45: Maximal input phase current under unbalanced voltage supply ($L = 10$ mH, $R = 0.1$ Ω, $V_{dc} = 600$ V).

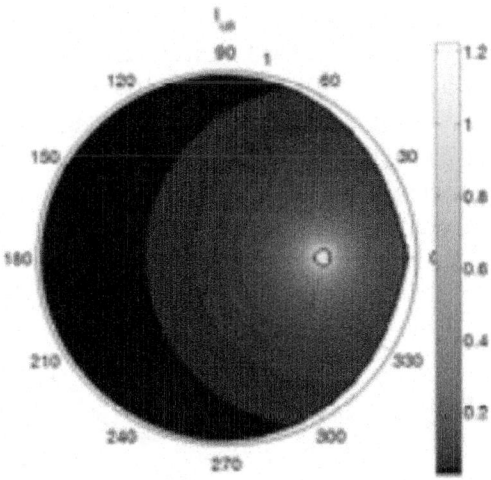

Figure 46: Input current unbalance under unbalanced voltage supply ($L = 10$ mH, $R = 0.1$ Ω, $V_{dc} = 600$ V).

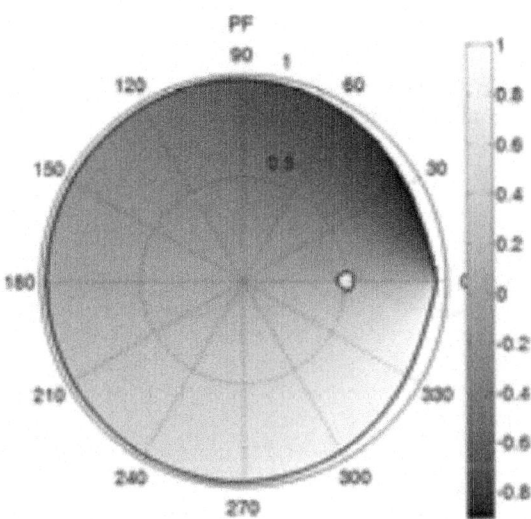

Figure 47: Power factor under unbalanced voltage supply ($L = 10$ mH, $R = 0.1$ Ω, $V_{dc} = 600$ V).

Measurements on an experimental system identical to the simulated one have been carried out in order to verify the investigated method. The scope traces in Figure 48 show the measured current in phase *A* and the DC link current when the negative-sequence in the supply voltage is not compensated for by the control method and the DC link current, therefore, contains significant component pulsating with a frequency of 100 Hz, twice the fundamental network frequency. The case when unbalanced voltage system is compensated by the investigated control method is illustrated in Figure 49. It can be seen that the pulsating component of the DC link current has been effectively eliminated by the investigated method.

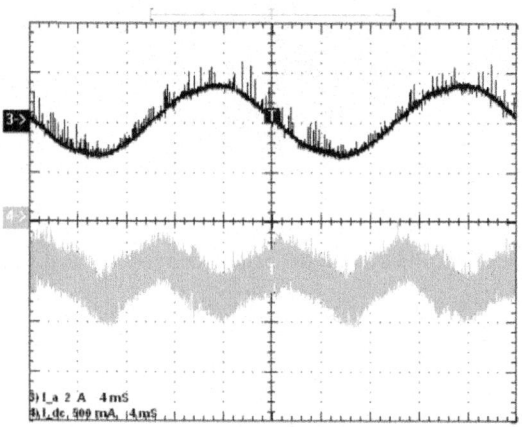

Figure 48: Phase A current and DC-link current under unbalanced voltage supply without elimination of pulsating component.

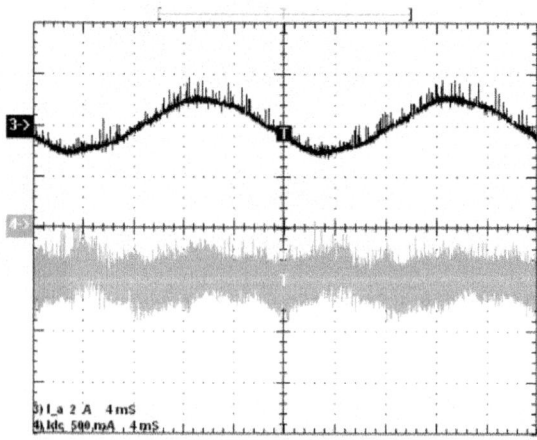

Figure 49: Phase A current and DC-link current under unbalanced voltage supply with elimination of pulsating component.

CONCLUSION

It has been shown in the article that it is possible to effectively compensate for the unbalanced voltage source at the input of a solid-state converter so that constant power flow into the DC bus is maintained. The results of simulations show that the choice of the operating point of front end converter may significantly affect the impact of the rectifier on the supplying power grid. It is possible to select the optimal operating point according to the chosen optimization criteria, which can be e.g. maximal power factor or current unbalance.

ACKNOWLEDGMENT

This work was supported by the Grant Agency of the Czech Republic under research grant No. 102/09/1273 and by the Institutional Research Plan AV0Z20570509.

REFERENCES

1. Stankovic, A. V. & Lipo, T. A. (2001). A Novel Control Method for Input Output Harmonic Elimination of the PWM Boost Type Rectifier Under Unbalanced Operating Conditions, IEEE Trans. on Power Electronics, 16, pp. 603-611, ISSN: 0885-8993.
2. Stankovic, A. V. & Lipo, T. A. (2001). A Generalized Control Method for Input-Output Harmonic Elimination of the PWM Boost Type Rectifier Under Simultaneous Unbalanced Input Voltages and Input Impedances, Power Electronics Specialists Conference, pp. 1309-1314, ISBN: 0-7803-7067-8, Vancouver, Canada, June 2001.
3. Lee, K.; Jahns, T. M.; Berkopec, W. E. & Lipo, T.A. (2006). Closed-form analysis of adjustable-speed drive performance under input-voltage unbalance and sag conditions, IEEE Trans.on Industry Applications, vol. 42, no. 3. pp. 733-741, ISSN: 0093-9994.
4. Cross, A. M.; Evans, P. D. & Forsyth, A. J. (1999). DC Link Current in PWM Inverters with Unbalanced and Non Linear Loads, IEE Proc.-Electr. Power Appl., vol. 146, no. 6, pp. 620-626, ISSN: 1350-2352.
5. Song, H. & Nam, K. (1999). Dual Current Control Scheme for PWM Converter Under Unbalanced Input Voltage Conditions, IEEE Trans. on Industrial Electronics, 46, pp. 953-959, ISSN: 0278-0046.
6. Chomat, M. & Schreier, L. (2005). Control Method for DC-Link Voltage Ripple Cancellation in Voltage Source Inverter under Unbalanced Three-Phase Voltage Supply, IEE Proceedings on Electric Power Applications, vol. 152, no. 3, pp. 494 – 500, ISSN: 1350-2352.

7. Chomat, M.; Schreier, L. & Bendl, J. (2007). Operation of Adjustable Speed Drives under Non Standard Supply Conditions, IEEE Industry Applications Conference/42th IAS Annual Meeting, pp. 262-267, ISBN: 978-1-4244-1259-4, New Orleans, USA, September 2007.
8. Chomat, M.; Schreier, L. & Bendl, J. (2009). Influence of Circuit Parameters on Operating Regions of PWM Rectifier Under Unbalanced Voltage Supply, IEEE International Electric Machines and Drives Conference, pp. 357-362, ISBN: 978-1-4244-4251-5, Miami, USA, May 2009.
9. Chomat, M.; Schreier, L. & Bendl, J. (2009). Operating Regions of PWM Rectifier under Unbalanced Voltage Supply, International Conference on Industrial Technology, pp. 510 – 515, ISBN: 978-1-4244-3506-7, Gippsland, Australia, February 2009.

CITATION

CHAPTER 1
M. Benhaddadi, G. Olivier, R. Ibtiouen, J. Yelle, and J-F Tremblay (2011). Premium Efficiency Motors, Electric Machines and Drives, Dr. Miroslav Chomat (Ed.), ISBN: 978-953-307-548-8, InTech, DOI: 10.5772/14893.

CHAPTER 2
ArieShenkman,SaadTapuchi,DmitryBaimel, (2015) A New Type of Capacitive Machine. Energy and Power Engineering,07,31-40. doi: 10.4236/epe.2015.72003

CHAPTER 3
Y. YANG, S. YANG and J. LIU, "Optimal Design and Control of a Torque Motor for Machine Tools," Journal of Electromagnetic Analysis and Applications, Vol. 1 No. 4, 2009, pp. 220-228. doi: 10.4236/jemaa.2009.14033.

CHAPTER 4
Nicolae D.V (2011). Electric Motor Performance Improvement Using Auxiliary Windings and Capacitance Injection, Electric Machines and Drives, Dr. Miroslav Chomat (Ed.), ISBN: 978 -953-307- 548 -8, InTech,

CHAPTER 5
Jalal Nazarzadeh and Vahid Naeini (2011). Magnetic Reluctance Method for Dynamical Modeling of Squirrel Cage Induction Machines, Electric Machines and Drives, Dr. Miroslav Chomat (Ed.), ISBN: 978-953- 307-548-8, InTech,

CHAPTER 6
Waldiberto de Lima Pires, Hugo Gustavo Gomez Mello, Sebastião Lauro Nau and Alexandre Postól Sobrinho (2011). Minimization of Losses in Converter-Fed Induction Motors – Optimal Flux Solution, Electric Machines and Drives, Dr. Miroslav Chomat (Ed.), ISBN: 978-953-307 -548-8, InTech,

CHAPTER 7
Jogendra Singh Thongam and Rachid Beguenane (2011). Sensorless Vector Control of Induction Motor Drive - A Model Based Approach, Electric Machines and Drives, Dr. Miroslav Chomat (Ed.), ISBN: 978-953-307-548- 8, InTech,

CHAPTER 8
Cristiane Cauduro Gastaldini, Rodrigo Zelir Azzolin, Rodrigo Padilha Vieira and Hilton Abílio Gründling (2011). Feedback Linearization of Speed-Sensorless Induction Motor Control with Torque Compensation, Electric Machines and Drives, Dr. Miroslav Chomat (Ed.), ISBN: 978-953-307-548-8, InTech,

CHAPTER 9
Rodrigo Z. Azzolin, Cristiane C. Gastaldini, Rodrigo P. Vieira and Hilton A. Gründling (2011). A RMRAC Parameter Identification Algorithm Applied to Induction Machines, Electric Machines and Drives, Dr. Miroslav Chomat (Ed.), ISBN: 978-953-307-548-8, InTech,

CHAPTER 10
Omar Hegazy, Amr Amin, and Joeri Van Mierlo (2011). Swarm Intelligence Based Controller for Electric Machines and Hybrid Electric Vehicles Applications, Electric Machines and Drives, Dr. Miroslav Chomat (Ed.), ISBN: 978-953-307-548-8, InTech,

CHAPTER 11
Miroslav Chomat (2011). Operation of Active Front -End Rectifier in Electric Drive under Unbalanced Voltage Supply, Electric Machines and Drives, Dr. Miroslav Chomat (Ed.), ISBN: 978-953-307-548-8, InTech,

INDEX

A

Adjustable speed drive (ASD) 10
Alternative current (AC) 34

B

Broken rotor 89, 104

D

Direct current (DC) 34

E

Electrical dynamic 54
Electrical machine 31
Electric power mechanisms 32
Electromechanical energy 89
Electromotive force (EMF) 46
Energy point 32

F

Feedback linearization control (FLC) 161
Field-oriented control (FOC) 181
Field-oriented controller (FOC) 202
Finite element method (FEM) 90
Fuel cell/supercapacitor hybrid electric vehicles (FCHEV) 236

H

High electric 32, 33, 42

I

Induction motor (IM) 161
Institute of electrical and electronics engineers (IEEE) 5
International electrotechnical commission (IEC) 9

L

Linearization modelling 162

M

Machines fault 89
Magnetic equivalent circuit method (MECM) 90
Magnetic flux 32, 111, 113, 117, 121, 126, 127
Magnetic material 112, 113
Main winding 68, 74, 75, 76, 79
Minimum energy performance standards (MEPS) 6
Model reference adaptive system

(MRAS) 162
Model reference control (MRC) 183, 187
Motor loss 112

P

Particles swarm optimization (PSO) 236
Particle swarm optimization (PSO) 201, 202, 208, 237
Permanent magnet (PM) 45
Power factor 67, 73, 76, 77, 78, 79, 80, 82, 83, 85, 86, 87
Power network 239, 240

R

Reluctance motor 67
Robust model reference adaptive controller (RMRAC) 182
Rotor circuit 67
Rotor flux 132, 133, 134, 135, 141, 142, 147, 148, 149, 150, 157, 158, 160
Rotor position 132
Rotor speed 132

S

Saturation 132
Single-phase induction motors (SPIM) 181
Suitable control 240
Swarm intelligence (SI) 236

T

Torque motor 45, 46, 47, 48, 54, 57, 61
Two-phase induction motor (TPIM) 203

U

Unsymmetrical voltages 239

V

Voltage drop 114

W

Wire size 68

Z

Zero-sequence current 241